Gene Targeting

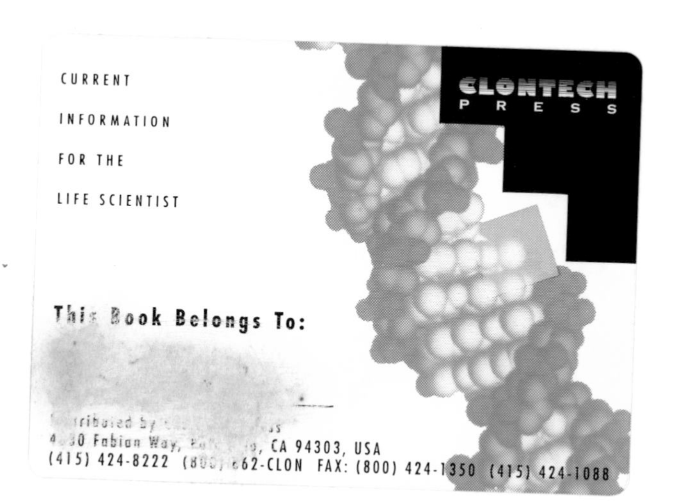

The Practical Approach Series

Affinity Chromatography
Anaerobic Microbiology
Animal Cell Culture (2nd Edition)
Animal Virus Pathogenesis
Antibodies I and II
Behavioural Neuroscience
Biochemical Toxicology
Biological Data Analysis
Biological Membranes
Biomechanics—Materials
Biomechanics—Structures and Systems
Biosensors
Carbohydrate Analysis
Cell–Cell Interactions
Cell Growth and Division
Cellular Calcium
Cellular Neurobiology
Centrifugation (2nd Edition)
Clinical Immunology
Computers in Microbiology
Crystallization of Nucleic Acids and Proteins
Cytokines
The Cytoskeleton
Diagnostic Molecular Pathology I and II
Directed Mutagenesis
DNA Cloning I, II, and III
Drosophila
Electron Microscopy in Biology
Electron Microscopy in Molecular Biology
Electrophysiology
Enzyme Assays
Essential Molecular Biology I and II
Experimental Neuroanatomy
Fermentation
Flow Cytometry
Gel Electrophoresis of Nucleic Acids (2nd Edition)
Gel Electrophoresis of Proteins (2nd Edition)
Gene Targeting
Gene Transcription
Genome Analysis
Glycobiology
Growth Factors

Haemopoiesis
Histocompatibility Testing
HPLC of Macromolecules
HPLC of Small Molecules
Human Cytogenetics I and II (2nd Edition)
Human Genetic Disease Analysis
Immobilised Cells and Enzymes
Immunocytochemistry
In Situ Hybridization
Iodinated Density Gradient Media
Light Microscopy in Biology
Lipid Analysis
Lipid Modification of Proteins
Lipoprotein Analysis
Liposomes
Lymphocytes
Mammalian Cell Biotechnology
Mammalian Development
Medical Bacteriology
Medical Mycology
Microcomputers in Biochemistry
Microcomputers in Biology
Microcomputers in Physiology
Mitochondria
Molecular Genetic Analysis of Populations
Molecular Neurobiology
Molecular Plant Pathology I and II
Molecular Virology
Monitoring Neuronal Activity
Mutagenicity Testing
Neural Transplantation
Neurochemistry
Neuronal Cell Lines
Nucleic Acid and Protein Sequence Analysis
Nucleic Acid Hybridisation
Nucleic Acids Sequencing
Oligonucleotides and Analogues
Oligonucleotide Synthesis
PCR
Peptide Hormone Action
Peptide Hormone Secretion
Photosynthesis: Energy Transduction
Plant Cell Culture
Plant Molecular Biology
Plasmids
Pollination Ecology
Postimplantation Mammalian Embryos
Preparative Centrifugation
Prostaglandins and Related Substances
Protein Architecture
Protein Engineering
Protein Function
Protein Phosphorylation
Protein Purification Applications
Protein Purification Methods
Protein Sequencing
Protein Structure
Protein Targeting
Proteolytic Enzymes
Radioisotopes in Biology
Receptor Biochemistry
Receptor–Effector Coupling

Receptor–Ligand Interactions
Ribosomes and Protein Synthesis
Signal Transduction
Solid Phase Peptide Synthesis
Spectrophotometry and Spectrofluorimetry
Steroid Hormones
Teratocarcinomas and Embryonic Stem Cells
Transcription Factors
Transcription and Translation
Tumour Immunobiology
Virology
Yeast

Gene Targeting

A Practical Approach

Edited by

ALEXANDRA L. JOYNER

Samuel Lunenfeld Research Institute
Mount Sinai Hospital and Department of Molecular and Medical Genetics
University of Toronto
Toronto, Canada

at
OXFORD UNIVERSITY PRESS
Oxford New York Tokyo

Oxford University Press, Walton Street, Oxford OX2 6DP
Oxford New York Toronto
Delhi Bombay Calcutta Madras Karachi
Kuala Lumpur Singapore Hong Kong Tokyo
Nairobi Dar es Salaam Cape Town
Melbourne Auckland Madrid
and associated companies in
Berlin Ibadan

Oxford is a trade mark of Oxford University Press

A Practical Approach is a registered trade mark
of the Chancellor, Masters, and Scholars of the University of Oxford
trading as Oxford University Press

Published in the United States
by Oxford University Press Inc., New York

A catalogue record for this book is available from the British Library

Library of Congress Cataloging in Publication Data
Gene targeting : a practical approach / edited by Alexandra L. Joyner.
(The Practical approach series)
Includes bibliographical references and index.
1. Genetic vectors. 2. Embryonic stem cells. 3. Genetic engineering. 4. Transgenic mice. I. Joyner, Alexandra L.
II. Series.
QH442.G4385 1993 599.32'33—dc20 92–44663
ISBN 0–19–963407–6 (hbk.)
ISBN 0–19–963406–8 (pbk.)

Typeset by Footnote Graphics, Warminster, Wilts
Printed in Great Britain by Information Press Ltd, Eynsham, Oxon

Preface

Our ability to mutate the mouse germline and to assay the consequences of such tinkering has expanded tremendously over the past five years due to imaginative uses of embryonic stem (ES) cells and to the application of homologous recombination to alter mammalian genes. ES cell lines are probably the most remarkable cell lines ever established since they can be cultured and manipulated relatively easily *in vitro* without losing their potential to step right back into their normal developmental programme when returned to the embryo. Homologous recombination when applied to altering specific endogenous genes, referred to as gene targeting, provides the highest possible level of control over producing mutations in cloned genes. Recently, the major barriers to efficiently producing targeted germline mutations in mice have been surpassed. With these new opportunities to apply genetic approaches in the mouse many workers have been attracted to the mouse to address questions of *in vivo* gene function. To date, gene targeting in ES cells has been used primarily to make null mutations for testing where and when in development a gene is required. The range of experiments will undoubtedly soon expand and gene targeting will be used, for example, for making gain-of-function and leaky mutations as well as to study gene regulation and protein structure/function relationships. ES cells have also proven useful for other applications, such as enhancer and gene trap screens where one can gain information on gene sequence, expression, and mutant phenotype all from a single insertion in ES cells.

The aim of this book is to describe in detail the rationale behind designing gene targeting vectors as well as protocols for the maintenance and many common manipulations of ES cells used for producing mutant mice. These include the isolation of genetically altered clones following electroporation, production of chimeras, and application of enhancer and gene trap strategies for identifying developmental genes. Two chapters on the production of ES cell chimeras were included since this is currently an area of experimentation. Although at present blastocyst injection is primarily used, aggregation chimeras have many advantages, such as their relative ease of production and the opportunity to produce complete ES cell-derived embryos, that make it an attractive alternative. A chapter also describes the latest techniques for culturing, manipulating, and assaying mouse and human bone marrow stem cells. This was included as a look toward the future, as it is highly likely that gene targeting will be used more and more to mutate genes in cell lines and correct mutant loci in tissue explants.

My hope is that this book can be used on its own as a manual for newcomers into the excitement and frustration of working with established ES cell lines to produce mutant mice. In addition, many of the protocols and strategies can easily be adapted to applications in cells other than ES cells. I would like to extend many thanks to all of the authors for their efforts to include such detailed protocols and lucid discussions of the various techniques presented. Finally, since many of the techniques described use animals or human subjects, the experiments should be carried out in accordance with local regulations.

Toronto, Canada A.L.J.
October 1992

Contents

Contributors

JOHN W. BELMONT
Institute for Molecular Genetics, 1 Baylor Plaza, Baylor College of Medicine, Texas Medical Center, Houston, Texas 77030, USA.

ALLAN BRADLEY
Institute for Molecular Genetics, 1 Baylor Plaza, Baylor College of Medicine, Texas Medical Center, Houston, TX 77030, USA.

ACHIM GOSSLER
Max Delbruck Laboratorium in der MPG, Carl-von-Linne Laboratorium, D-5000 Koln 30, Germany.

PAUL HASTY
Institute for Molecular Genetics, 1 Baylor Plaza, Baylor College of Medicine, Texas Medical Center, Houston, TX 77030, USA.

RANDALL JOHNSON
Hormone Research Institute, Department of Biochemistry and Biophysics, UCSF, San Francisco, CA 94143, USA.

ALEXANDRA JOYNER
Samuel Lunenfeld Research Institute, Mount Sinai Hospital, 600 University Avenue, Toronto, Canada M5G 1X5.

KATERI A. MOORE
Institute for Molecular Genetics, 1 Baylor Plaza, Baylor College of Medicine, Texas Medical Center, Houston, TX 77030, USA.

ANDRAS NAGY
Samuel Lunenfeld Research Institute, Mount Sinai Hospital, 600 University Avenue, Toronto, Canada M5G 1X5.

VIRGINIA PAPAIOANNOU
Department of Pathology, Tufts University, 136 Harrison Avenue, Boston, MA 02111, USA.

JANET ROSSANT
Samuel Lunenfeld Research Institute, Mount Sinai Hospital, 600 University Avenue, Toronto, Canada M5G 1X5.

WOLFGANG WURST
Samuel Lunenfeld Research Institute, Mount Sinai Hospital, 600 University Avenue, Toronto, Canada M5G 1X5.

JOCHEN ZACHGO
Max Delbruck Laboratorium in der MPG, Carl-von-Linne Laboratorium, D-5000 Koln 30, Germany.

Abbreviations

BCS	bovine calf serum
BFU	burst-forming unit
BMC	bone marrow cell
BRL	Buffalo rat liver
BSA	bovine serum albumin
CFC	colony-forming cells
c.f.u.	colony-forming units
CFU-G	colony-forming unit granulocyte
CFU-GM	colony-forming unit granulocyte/macrophage
CFU-M	colony-forming unit macrophage
CFU-S	colony-forming unit spleen
CSF	colony-stimulating factor
ddH_2O	double distilled water
DMEM	Dulbecco's modified Eagle's media
DMSO	dimethyl sulfoxide
dpc	days post-coitus
DT	diphtheria toxin
EDTA	ethylenediamine tetra-acetic acid
EMFI	primary embryonic fibroblasts
ES	embryonic stem
ET	enhancer trap
FACS	fluorescence activated cell sorting
FCS	fetal calf serum
FDG	fluorescein diβ-D galactopyranoside
G418	geneticin^{tm}, Gibco
GDA	glutaraldehyde
GPI	glucose phosphate isomerase
gpt	*guanine phosphoribosyl transferase*
GT	gene trap
hCG	human chorionic gonadotropin
hisD	*histidinol dehydrogenase*
HPP	high proliferative potential
hprt	*hypoxanthine-guanine phosphoribosyl transferase*
HSC	heamopoietic stem cells
HSV	herpes simplex virus
ICM	inner cell mass
IMDM	Iscove's modified Dulbecco's media
IU	international units
IV	intravenous

kb	kilobase
LH	luteinizing hormone
LIF	leukaemia inhibitory factor
LTC-IC	long-term culture-initiating cells
LTR	long terminal repeat
MC	methylcellulose
m.o.i.	multiplicity of infection
Mo-MuLV	Moloney murine leukaemia virus
MRA	marrow repopulating ability
MTG	monothioglycerol
neo	*neomycin*
neo^R	*neomycin* resistance
PBS	phosphate-buffered saline
PCR	polymerase chain reaction
PFA	paraformaldehyde
PHA-LCM	phytohaemagglutinin-stimulated lymphocyte conditioned medium
PMSG	pregnant mare serum gonadotropin
PT	promoter trap
RACE	rapid amplification of cDNA ends
RCV	replication-competent virus
rt	room temperature
SBA	soybean agglutinin
SDS	sodium dodecyl sulphate
SPF	specific pathogen free
TBI	total body irradiance
tk	*thymidine kinase*
TM	melting temperature
ZP	zona pellucida

1

Gene targeting vectors for mammalian cells

PAUL HASTY and ALLAN BRADLEY

1. Introduction

When a fragment of genomic DNA is introduced into a mammalian cell it can locate and recombine with the endogenous homologous sequences. This type of homologous recombination, known as gene targeting, is the subject of this chapter. Gene targeting has been widely used, particularly in mouse embryonic stem (ES) cells, to make a variety of mutations in many different loci so that the phenotypic consequences of specific genetic modifications may be assessed in the organism.

In mammalian cells the first experimental evidence for the occurrence of gene targeting was made using a fibroblast cell line with a selectable artificial locus by Lin *et al.* (1), and was subsequently demonstrated to occur at the endogenous β-globin gene by Smithies *et al.* in erythroleukaemia cells (2). In general, the frequencies of gene targeting in mammalian cells are relatively low and this is probably related, at least in part, to the competing pathway: the fact that transfected DNA can also integrate into a random chromosomal site. The relative frequency of targeted to random integration events will determine the ease with which targeted clones may be identified in a gene targeting experiment. This chapter details a variety of aspects of vector design which can determine the type of mutation that may be generated in the target locus, as well as the selection and screening strategies which can be used to identify clones of cells with the desired targeted modification. Since the most common experimental strategy, at least in the first instance, is to ablate the function of a target gene by introducing a selectable marker gene, we initially describe the vectors and the selection schemes which are helpful in the identification of such recombinant clones (Sections 2–5). The vectors and additional considerations for generating mutations in a target locus which do not include selectable markers are also discussed (Section 6).

1.1 Targeting vectors

A targeting vector is designed to recombine with and mutate a specific chromosomal locus. The minimal components of such a vector are homologous sequences to the desired chromosomal integration site and a plasmid backbone. Since both the transfection and targeting frequency of such a vector can be low, it is desirable to include other components in the vector such as positive and negative selection markers which provide strong selections for the targeted recombination product (Sections 4 and 5). The positive selection marker in a targeting vector may serve two functions. Its primary purpose is as a selection marker to isolate the rare transfected cells that have integrated DNA (which occur at a frequency of about one in 10^4). Secondly, the positive selection marker can serve as a mutagen, for instance if it is cloned into a coding exon of the gene or replaces coding exons.

Two distinct vector designs are commonly used for targeting in mammalian cells, replacement and insertion vectors (*Figure 1*). These vector-types are configured differently so that following homologous recombination they yield different integration products. For the purpose of clarity, aspects of vector design which are relevant to both replacement and insertion vectors, such as the length of homology, selection cassettes, and enrichment schemes will be discussed in Section 4.5. However, there are a number of unique considerations for both vector-types and these are discussed in Sections 2 and 3.

A Replacement Vector

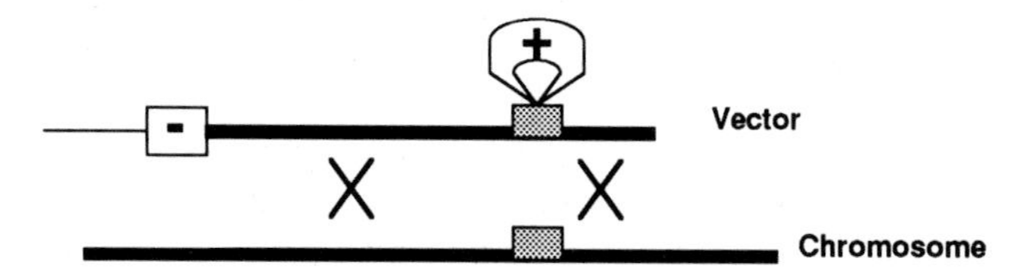

B Insertion Vector

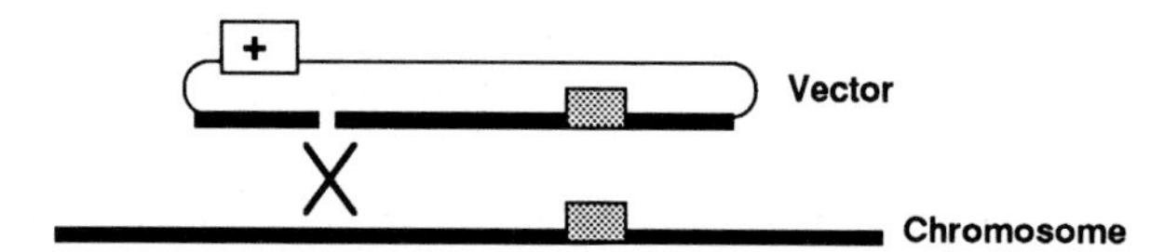

Figure 1. Diagram of a replacement and insertion vector. The thick line represents the vector homology to the target locus; the thin line represents bacterial plasmid. The stippled rectangle represents an exon. The positive selection marker is shown as a box that contains a +. (A) The replacement vector. The positive selection marker interrupts the target homology. This is required for a replacement vector. The negative selection marker is shown as a rectangle that contains a −. The replacement vector is linearized outside the target homology prior to transfection. (B) An insertion vector. A positive selectable marker may be cloned into the homologous sequences or the vector backbone. A double strand break is generated in the target homology prior to transfection.

2. Replacement vectors

The fundamental elements of a replacement vector are the homology to the target locus (see Section 4.1), a positive selection marker (see Sections 4.2–4.5 and 5), bacterial plasmid sequences, and a linearization site outside of the homologous sequences of the vector. In some cases a negative selection marker may also be used to enrich the transfected cells against random integration events (Section 4.5 and 5). The basic design of a replacement vector is illustrated in *Figure 1A*. The mechanistic details of recombination pathways used by replacement vectors are beyond the scope of this chapter, but the final recombinant allele can be effectively described as a consequence of double reciprocal recombination which takes place between the vector and the chromosomal sequences (*Figure 2A*). The final recovered product is equivalent to a replacement of the chromosomal homology with all components of the vector which are flanked on both sides by homologous sequences; any heterologous sequences at the ends of the vector homology are excised from the vector and are not recovered as stable genomic sequences following targeting. This latter feature has been used as a basis to enrich populations of transfected cells for targeted integration events (Section 4.5).

2.1 Design considerations of a replacement vector

The principal consideration in the design of a replacement vector, is the type of mutation generated. Secondary (yet still important) considerations relate to the selection and screening techniques required to isolate the recombinant clones. The recombinant alleles generated by gene targeting experiments typically have a selection cassette inserted into a coding exon. However, this will not necessarily ablate the function of the target gene to generate a null allele. Consequently, it is necessary to confirm that the allele which has been generated is null by RNA and/or protein analysis and in many cases transcripts and truncated proteins from such a mutant allele can be detected. Considering that products from the mutated locus may have some function (normal or abnormal) it is important to design a replacement vector so that the targeted allele which is generated is null, particularly in the absence of a good assay for the gene product.

The primary variable that can influence the type of mutation which is generated with a replacement vector is the location and orientation of the positive selection marker in the target gene. Disruption of the coding sequence by the positive selection marker will in most instances ablate a gene's function. However, in some situations a truncated protein may be generated which retains some biological activity, thus some knowledge of mutations in other systems will be helpful in the determination of the possible function of a targeted allele. Null alleles are more likely to occur by recombining a selection cassette into an upstream exon rather than a downstream exon, since under these circumstances minimal portions of the wild-type polypeptide should be made.

In most cases, the positive selection marker should be inserted into an exon. Since the length of an exon can influence RNA splicing (3), an artificially large exon with an inserted selectable marker may not be recognized by the splicing mechanism and could be skipped. Thus, transcripts initiated from the endogenous promoter may delete the mutated exon from the mRNA species. If this involves a coding exon without a unit number of codons, the net result will be both a deletion and a frame-shift mutation of the gene, which should generate a null allele. However, if the disrupted coding exon has a unit number of codons which is spliced out, this would result in a protein with a small in-frame deletion which may retain partial or complete function.

From the preceding discussion it is clear that the choice of how to effectively mutate a locus is complex. For most purposes it is advisable to delete portions or all of the target gene so that the genetic consequences are not ambiguous. Such a deletion may be accomplished by using the positive selectable marker to replace gene sequences. Since comparative data is not yet available to address the frequency of recombination for vectors in which sequences have been deleted, a conservative vector design would encompass a deletion which would eliminate no more than three kilobases, even though larger deletions have been reported (4). It is important to be aware that deletions, particularly if they are large, may effect multiple genes in situations where they are located adjacent or internal to the target locus.

The length of the homologous sequences will affect the targeting frequency (see Section 4.1). For most vectors the length of homology should be in the range of five to eight kilobases. The position of the positive selection marker with respect to the homologous sequences of the vector will determine the types of screens that must be used to find the clones targeted with a gene replacement event. One common screening tool for targeted clones is based on the polymerase chain reaction (PCR) which can be designed to detect the juxtaposition of the vector and the target locus (5, 6). The use of PCR to detect targeted clones is described in detail in Chapter 2. This is accomplished by using one primer which anneals to the positive selection marker in the targeting vector and a second primer which anneals to the target chromosomal sequences just beyond the homology used in the vector (*Figure 2A*). The efficiency of such a PCR amplification is related to the distance between the unique primer site in the vector (usually the positive selection marker) and the sequence external to the homologous elements of the vector which should be in the 500 base pair to two kilobase range. Thus, replacement vectors configured for screens by PCR require the positive selection marker to be inserted at an asymmetric location near one end of the homologous sequences, while still leaving sufficient homology for the formation of a cross-over. This will give vectors with one long arm and one short arm of homologous sequences (*Figure 2A*).

Another common screen for clones targeted with gene replacement vectors uses Southern blot analysis. For this it is important to design the vector and identify unique probes and restriction sites so that such analysis is both unambiguous and can discriminate the various categories of recombinant clones (Section 2.3).

Since a replacement vector should be linearized before transfection into cells at a site outside the homologous sequences, the cloning steps must incorporate at least one unique restriction enzyme site outside the homologous sequences. There is no advantage in releasing the homologous sequences of the vector from the bacterial plasmid (7).

The following are guidelines for the construction of a replacement vector which generates an easily identifiable null allele.

(a) Use a fragment of homologous sequences of 5–8 kb.

(b) Insert the positive selectable marker into an upstream exon.

(c) Interrupt an internal exon without a unit number of codons to avoid generating a protein product with partial function due to a deletion of the mutated exon by RNA splicing.

(d) Delete the entire gene if it is small, or important 5′ exons if it is large.

(e) If PCR is to be used to screen for gene replacement events, clone the positive selection marker so that one arm of homology is 0.5–2 kb.

(f) Linearize the vector outside the homologous sequences.

Many replacement vectors are used with a negative selection marker at the end(s) of the vector which can aid in the enrichment for targeted events (Section 4.5). The vector may contain two selection cassettes (one at each end), although in this instance it is advisable to minimize the chance for the occurrence of extrachromosomal intermolecular homologous recombination. This may be accomplished by using different genes or negative selection cassettes which have diverged at the sequence level such as *HSV1tk* and *HSV2tk*.

2.2 Recombinant alleles generated by replacement vectors

The desired genetic exchange of a replacement vector is one in which vector sequences effectively replace the homologous region in the genome (*Figure 2A*). However, undesirable targeted products occur with replacement vectors (7); primarily with vectors whose ends have been joined to form concatemers, circles, or both (*Figure 2B*). The final recombination product of a concatemer or a circle is the integration of the entire vector including the bacterial plasmid and other associated heterologous sequences. If the vector was intact at the time of the ligation event many of these undesirable products will be eliminated by negative selection because they incorporate the entire vector

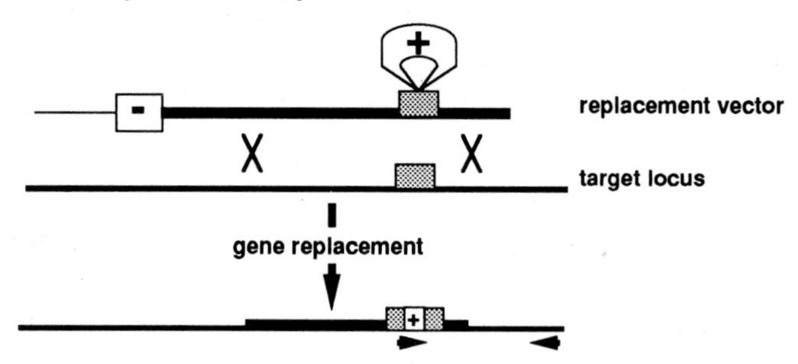

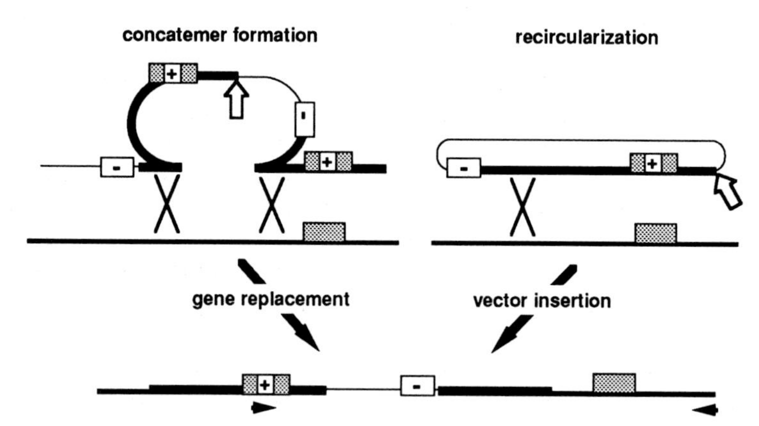

Figure 2. Integration patterns for a replacement vector. The replacement vector is linearized outside the target homology prior to transfection. The thick line represents the vector homology to the target locus; the thin line represents bacterial plasmid. The line of intermediate thickness represents the target locus in the chromosome. The stippled rectangle represents an exon. The positive selection marker is shown as a box that contains a +. The negative selection marker is shown as a rectangle that contains a −. The Xs represent cross-over points. The arrow heads represent PCR primers, one in the positive selection marker and the other in the target locus adjacent to the target homology. (A) Gene targeting that results in a simple gene replacement product and a PCR junction fragment. (B) Gene targeting that results from the integration of all vector components. The open arrows point to the location for extrachromosomal end-to-end joining that concatenates or recircularizes the vector. The PCR junction fragment is too large to be amplified.

including the negative selection marker. These undesirable products will also not be scored as positive by most PCR screens for targeted events since these screens are usually specific for the recombination events that have crossed-over on the short arm of the vector while most of the alternative events are the products of recombination events which occur through the long arm of the vector.

2.3 Replacement vectors—screening for targeted events

One of the most important aspects of any gene targeting experiment is to confirm that the desired genetic change has occurred (*Figure 3*). Given that replacement vectors may integrate via a number of different pathways and result in different targeted mutations, it is important to resolve the different classes of integration events, particularly gene replacement events that introduce only a positive selection marker from recombination events which result from integration of the entire vector or concatemers. To clarify the following discussion, targeted clones which have only incorporated the positive selection marker are referred to as simple gene replacement events.

Following the transfection of a replacement vector, colonies which survive positive and negative selection should be clonally isolated and screened by either PCR or Southern blot analysis for a specific recombination allele (see Chapter 2). The screen should be designed to find targeted events rapidly, generally it is desirable to make this screen specific for the targeted clones which have recombined with the vector on the short arm of homology because these tend to be more diagnostic for a simple gene replacement event. For PCR-based screens the primers are chosen so that they will amplify a specific junction fragment following a cross-over on the short arm of the vector (5, 6). If the PCR amplification is efficient the PCR product may be visually detected on an ethidium bromide stained gel. Occasionally it may be necessary to hybridize the PCR product to a labelled oligonucleotide probe (specific for sequences outside the vector) to confirm the identity of the amplification product. Southern blot-based screens are also possible at an early stage in the expansion of the transfected clones. These screens may be performed on the small amounts of DNA obtained from cells grown in 96 or 24 well plates (reference 8, see Chapter 2 for details); for Southern blot analysis the choice of restriction digest and probe must readily distinguish the wild-type from the predicted targeted allele. Following a primary screen, the recombinant allele present in the putative positive clone must be subjected to extensive restriction analysis to confirm its structure. This analysis is ideally performed with a probe that is not contained in the target vector (external probe) and a restriction digest with an enzyme that does not cut in the vector (*Figure 3A*). Under these circumstances a simple gene replacement event will increase the length of the wild-type fragment by the size of the positive selection marker. The integration of the entire vector will be readily detectable by a very large increase in fragment size. Although this is the ideal, it is often difficult to find a restriction enzyme that does not cut in the vector while still generating genomic fragments that can be readily resolved on a gel.

The integration of the entire vector or concatemers of the vector may be difficult to discriminate from simple replacement events if the restriction digest detects a site within the vector and the probe is from outside the vector (*Figure 3B*). This type of Southern analysis can only analyse one arm of the

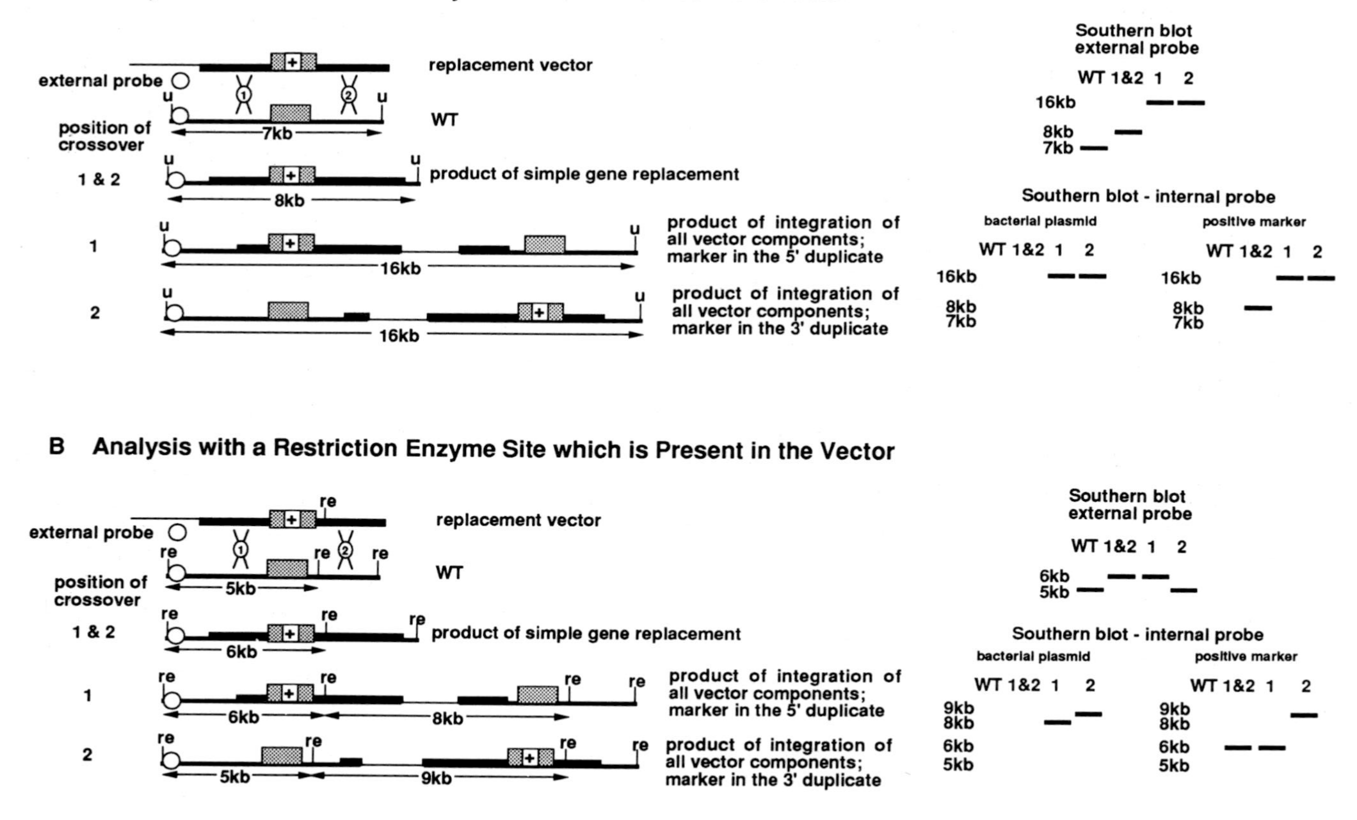
A Analysis with a Restriction Enzyme Site which is Not in the Vector
external probe
replacement vector
position of crossover
WT
7kb
1 & 2
product of simple gene replacement
8kb
1
product of integration of all vector components; marker in the 5' duplicate
16kb
2
product of integration of all vector components; marker in the 3' duplicate
16kb
Southern blot external probe
WT 1&2 1 2
16kb
8kb
7kb
Southern blot - internal probe
bacterial plasmid
WT 1&2 1 2
16kb
8kb
7kb
positive marker
WT 1&2 1 2
16kb
8kb
7kb
B Analysis with a Restriction Enzyme Site which is Present in the Vector
external probe
replacement vector
position of crossover
WT
5kb
1 & 2
product of simple gene replacement
6kb
1
product of integration of all vector components; marker in the 5' duplicate
6kb
8kb
2
product of integration of all vector components; marker in the 3' duplicate
5kb
9kb
Southern blot external probe
WT 1&2 1 2
6kb
5kb
Southern blot - internal probe
bacterial plasmid
WT 1&2 1 2
9kb
8kb
6kb
5kb
positive marker
WT 1&2 1 2
9kb
8kb
6kb
5kb

Figure 3. Southern blot analysis to differentiate simple gene replacement events from the integration of all vector components into the target locus. The lines and boxes are the same as in *Figure 2*. The first and second cross-over point is depicted by an X labelled either 1 or 2. The circle represents a probe external to the vector. The bacterial plasmid and the positive selection markers can be used as probes internal to the vector. The panels to the right are examples of Southern blots. WT, wild-type; 1 & 2, simple gene replacement event; 1, integration of the entire vector with the positive selection marker in the 5′ duplicate; 2, integration of the entire vector with the positive selection marker in the 3′ duplicate. (A) Southern analysis with a restriction enzyme, u, that does not cut in the vector. (B) Southern analysis with a restriction enzyme, re, that cuts once in the vector.

---

target locus, and will score the integration of the entire vector either as a simple gene replacement event or as a wild-type allele depending upon the location of the cross-over. It is therefore important to analyse both the 5′ and 3′ aspects of the target locus by using external probes for both ends. An internal probe (such as the positive selectable marker) should also be used since a simple gene replacement allele should give a different size fragment compared to an allele which is generated by the integration of the entire vector. Plasmid sequences may be used as probes since they will only detect recombinant alleles generated during vector insertion-like events.

One problem with internal probes is that they will also detect random integration events. This can result in misdiagnosis of an allele, if for instance a clone with a simple gene replacement allele and a separate random integration event is checked with an internal probe it may appear to have integrated all of the vector components into the target locus. To avoid these false negatives, it is advisable to perform Southern blot analysis with several restriction digests in which at least one of the restriction sites is outside the target vector. If this is done, predicted size fragments would only be consistently seen with simple gene replacement alleles whereas non-predicted fragments would be seen following random integration.

One class of integration event which may be detected in the primary screens for recombinant clones can appear to have been targeted but are not true targeted events (6). Mechanistically these involve a strand exchange of the vector with the target which results in the vector picking up some sequences which flank the target site. This homologous recombination intermediate is not resolved, consequently the vector dissociates from the target and may integrate at a random location to generate a partial third allele. Under these circumstances, the recombinant clones will score as positive by PCR and with some restriction digests by Southern analysis using flanking probes on one side. However, if the Southern analysis is extensive these types of clones may be identified because the generation of a novel targeted restriction fragment would not be accompanied by the usual reduction in intensity of the endogenous fragment.

3. Insertion vectors

The basic elements of an insertion vector are the same as those in a replacement vector (*Figure 1B*). The major difference between the two vector types is that the linearization site of an insertion vector is made in the homologous sequences of the vector.

An insertion vector undergoes single reciprocal recombination (vector insertion) with its homologous chromosomal target which is stimulated by a double strand break or gap in the vector (*Figure 4*). Our observations have indicated that an insertion vector can target at a 5–20-fold higher frequency than replacement vectors given the same homologous sequences (7). However if an insertion vector is configured so that there are severe topological constraints then this elevation in targeting frequency may not be noted (9). Since the entire insertion vector is integrated into the target site, including the homologous sequences of the vector, the recombinant allele generated by such a vector becomes a duplication of the target homology separated by the heterologous sequences in the vector backbone.

3.1 Vector design for insertion vectors

The essential components of an insertion vector are a region of homology to the target locus which has a unique linearization site, a positive selection marker (which serves to select for transfected cells in the same way as it would in a replacement vector), and a bacterial plasmid backbone. If the homologous sequences in the vector are wild-type (no mutations), then the position of the linearization point within the homologous sequences should not significantly affect either the frequency of the recombination event or the structure of the mutant allele. However, if one of the exons has been mutated, the position of the linearization site will affect the structure of the recombinant allele. These considerations are discussed in more detail in Section 6.3. Because the entire vector integrates into the target locus, the positive selection marker may be positioned in either the vector backbone or the homologous sequences of the vector. Since insertion vectors will generate a duplication of the homologous sequences in the vector, the position of the selection cassette may not play a critical role in the generation of a null mutation. However, if the selection cassette is cloned into the only exon in the vector then after vector insertion, the targeted allele will contain one normal and one artificially large exon. Exon skipping of this large exon may result in normal transcripts being generated from such a locus at a low frequency (10). A variety of transcripts may be produced due to exon skipping as demonstrated in *Figure 4C* in which the insertion vector contains one exon disrupted with the positive selection marker and one wild-type exon.

When designing an insertion vector it is important to consider the potential RNA species and coding possibilities of the recombinant allele. For most recombinant alleles generated with insertion vectors the duplication of exonic sequences is usually sufficient to mutate the gene, however, this will depend on a number of factors. For instance, a duplication of exonic sequence without a frame-shift mutation may not necessarily ablate protein function. To ensure the exon duplication will create a mutation at the protein level it may be necessary to introduce stop codons into a single exon in the targeting vector. A mutation created by frame-shifting in the carboxy-terminus of a protein may leave a functional N-terminal domain which may retain partial or complete activity. Exon duplications which are 5′ or 3′ to coding sequences or include the first or last coding exon will still have a collinear intact genomic sequence after the targeted integration and may be functionally wild-type (*Figure 4D*).

At the design stage consideration should be given to the screening procedure for the targeted allele. Because the genomic locus contains a duplication, the screen for an insertion event may be complex. These considerations are discussed in Section 3.2.

The basic design requirements of an insertion vector are as follows.

(a) Use 5–8 kb of homology.
(b) A unique restriction enzyme site is needed to generate a double strand break in the homology.
(c) A positive selection marker located either in the homologous sequences or in the backbone.
(d) If there is only one exon in the vector, do not disrupt it with the positive selection marker.
(e) Do not use a genomic fragment with the first or last coding exon in the vector.
(f) Destroy restriction sites for easy Southern blot analysis.

3.2 Screening for recombinant alleles generated with insertion vectors

In contrast to the different clases of integration events that may be generated with a replacement vector, an insertion vector usually integrates in a highly predictable way (unless there is a mutation in the homologous sequences). The first step in screening transfected cells for an allele generated with an insertion vector may be done by PCR or Southern blot analysis on clones selected for the presence of the positive selection marker. There are two ways to configure an insertion vector so that the recombinant locus can generate a specific PCR product. The most reliable PCR screen is based on the repair of DNA deleted from the insertion vector at the position of the linearization site to form a gap as shown in *Figure 4B* (11). Thus, PCR for the recombinant locus will use a primer that corresponds to the deleted sequence and a second primer which is specific for heterologous sequences in the vector. Upon

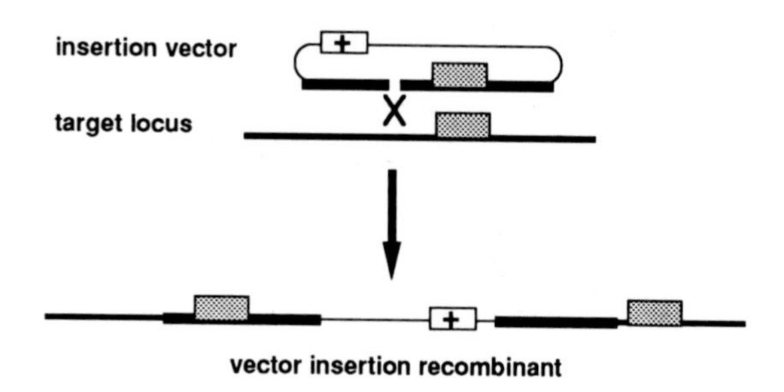

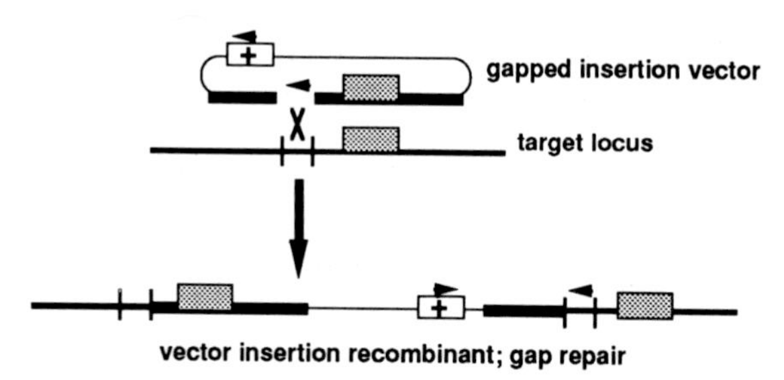

Figure 4. The integration pattern for an insertion vector. The insertion vector is linearized inside the target homology prior to transfection. The lines and boxes are as indicated in *Figure 2*. (A) An insertion vector with one exon, the positive selection marker should not be in the exon. (B) PCR screen for insertion events via gap repair. The insertion vector contains a gap at the linearization point, demarcated by two vertical lines in the target locus and the vector insertion recombinant. The arrow heads represent PCR primers; one is located in the positive selection marker and the other in the DNA which corresponds to the gap that is repaired during the targeting event. (C) An insertion vector with two exons, the positive selection marker disrupts exon 2. The arrow heads represent PCR primers; one is in the positive selection marker and the other is in the target locus immediately outside the target homology in the vector. The potential splicing products are shown under each target locus. (D) An insertion vector with the 5′ noncoding sequences present in the vector. The positive selection marker contains a promoter and is in the same transcriptional orientation as the target locus. The endogenous promoter and polyadenylation sequence are represented by an encircled pro and pA, respectively. The endogenous translation start is indicated by ATG. The arrows represent transcription: *top*, from the endogenous promoter or *bottom*, from the promoter in the positive selection marker to form a dicistronic messge. Dicistronic messages have been shown to result in the translation of both cistrons (34).

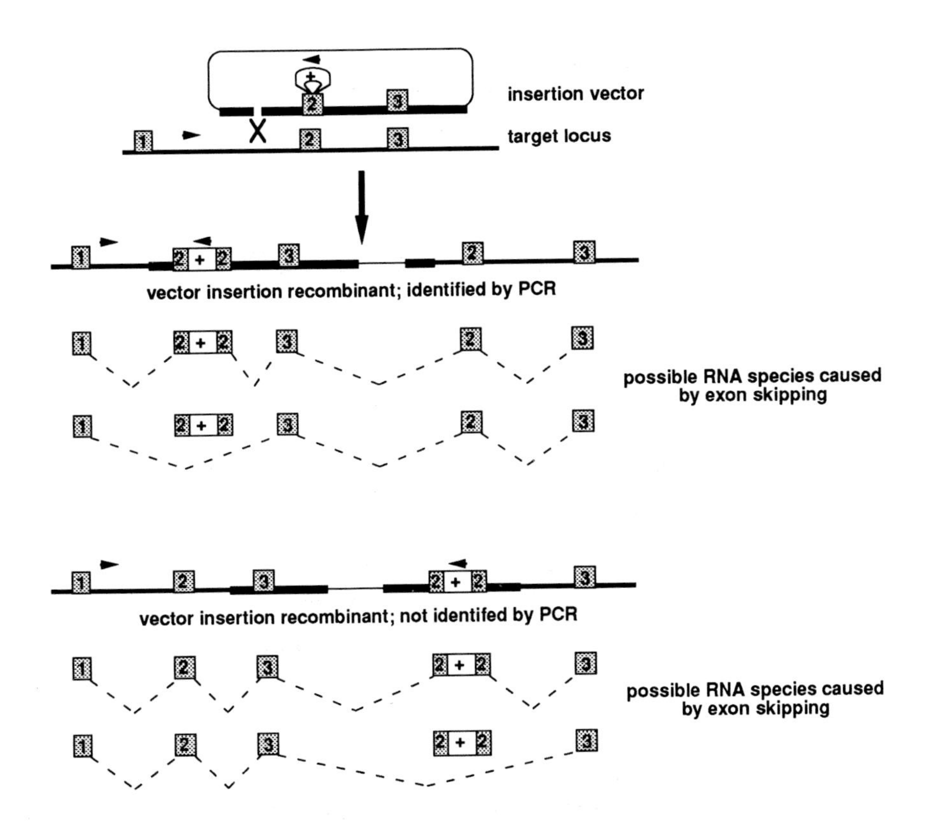
C Two Exons: Positive Selection Marker disrupts the Homology
insertion vector
target locus
vector insertion recombinant; identified by PCR
possible RNA species caused by exon skipping
vector insertion recombinant; not identifed by PCR
possible RNA species caused by exon skipping

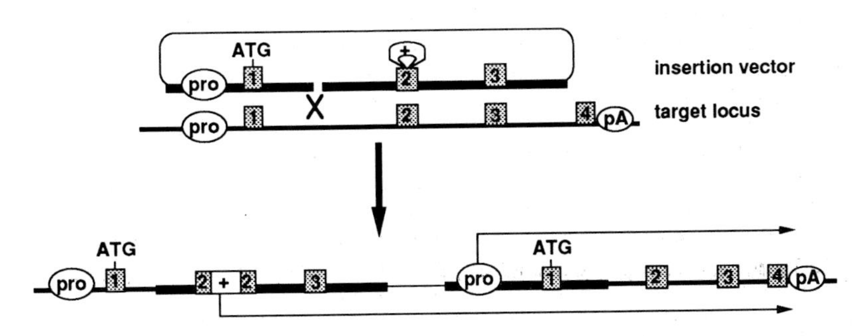
D An Insertion Vector with the 5' Noncoding sequences
ATG
pro
insertion vector
target locus
pA
vector insertion recombinant; complete, uninterrupted coding sequence present

insertion into the target locus, the gap will be repaired using the chromosomal sequences as a template. This will allow amplification by PCR, while insertions of the vector into random locations in the genome will only have the single primer site in the vector and thus will not provide an appropriate template for PCR amplification. A second way to design a PCR screen is to use a vector in which the homologous sequences have been identified with a mutation, this mutation can serve as a primer site for PCR coupled with a primer site external to the vector (*Figure 4C*). However, since the mutation in an insertion vector can end up on either side of the duplicated sequence, some targeted clones will not be identified by this assay (for reasons described in Section 6.3.1). Southern blot based screens using the 96 or 24 well protocol are also possible with insertion vectors (reference 8, Chapter 2). Probe(s) used for this preliminary screen should not be included in the homologous sequences of the vector.

After the PCR or Southern blot analysis, positive clones identified in the preliminary screen should be expanded and the integration pattern confirmed by detailed Southern blot analysis, preferably using a probe which is external to the target sequences. The best diagnostic digest for Southern blot analysis would be with an enzyme that does not cut within either the homology of the target locus or the vector backbone. A clone which has inserted one unit of the vector into the target locus will show an increase in the restriction fragment by the size of the vector. Concatemers of the vector integrated into the target locus may also be readily identified in the same assay. Since it may not be possible to identify a restriction enzyme that does not cut the vector or target locus to generate fragments which are readily resolvable (less than 20 kilobases) on an agarose gel, it is often necessary to use multiple digests as well as internal and external probes as described for replacement vectors (see Section 2.3). As stated previously, it may be possible to avoid problems in identifying unique restriction sites for Southern blot analysis by mutating sites in the vector or choosing a region of homology for vector construction which lacks specific enzyme sites. *Ideally* the Southern blot strategy and single copy probes should be defined before the vector is made.

4. Maximizing the targeting frequency and selection of targeted clones

A targeting vector transfected into cells is subject to two competing reactions, it may recombine into the desired chromosomal target, or more frequently the vector will integrate into a random site in the genome. It is thus desirable to design the vector to increase the chance of, and maximize selection for, clones with a targeted integration event and against clones with random insertion events. Since gene targeting is designed to precisely modify a chromosomal locus, it is possible to utilize aspects of the genetic exchange and elements of the recombinant locus to provide strong selections and enrichments for clones of cells with targeted integration events.

At any given locus there are two major variables which can influence the targeting frequency that may be controlled experimentally. First, the choice of an insertion or replacement vector can significantly affect the targeting frequency (7, 11). Secondly, the homologous sequences in the vector, specifically their length (9, 12) and the degree of polymorphic variation between the vector and the chromosome (13) have been shown to effect targeting frequencies.

4.1 Homology to the target locus

The best documented factor that effects the recombination frequency in mammalian cells is the length of homology to the target locus (*Figure 5*). A number of groups have described a relationship between the length of homology and targeting frequency (9, 12). As a general rule the greater the length of homology the higher the targeting frequency will become. For most vectors we recommend the ideal length of the homologous sequences in the vector to be in the range of five to ten kilobases. Although the use of even greater lengths of homology will probably increase the targeting frequency, the vectors will begin to get unwieldy due to both their physical size and the limited choice of unique restriction enzyme sites which can be used both to linearize the vector prior to transfection and later for detailed unambiguous Southern blot analysis of transfected clones.

As discussed previously, the positive selection cassette/mutation present in a replacement vector may divide the homology into a long and short arm. The experimental desire to distribute the homology asymmetrically is associated with the use of PCR to screen for specific junction fragments, since the efficiency of amplification by PCR is affected by the distance between the

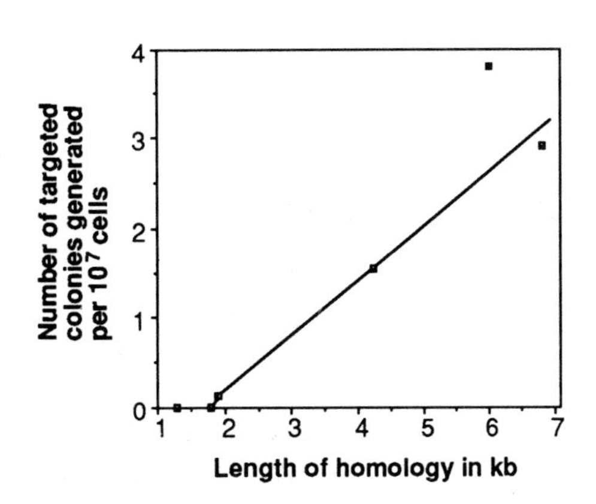

Figure 5. The relationship between the targeting frequency and the length of homology in a replacement vector (12).

primers. This asymmetry can have an effect on the targeting frequency since the length of the short arm may become suboptimal. Although cross-overs have been observed to occur with less than 500 base pairs on the short arm (12) it is probably desirable to work within the range of two kilobases for efficient pairing and cross-over formation (14). If PCR screens are not a part of the screening strategy then it may be an advantage to evenly distribute the homology on both sides of the positive selection marker.

Insertion vectors may also be described as having a long and short arm if a mutation disrupts the homologous sequence. If the double strand break is made too close to a large interruption of homology (such as a selectable marker) this can decrease the targeting frequency. The vector should be designed so that the double strand break can be made at least one and a half kilobases away from the selection cassette and the backbone. The location of the double strand break with respect to subtle mutations in the homologous sequences does not appear to greatly affect the targeting frequency, although this is an important consideration when using 'Hit and Run' vectors to make subtle mutations (see Section 6.3).

It is known from studies on extrachromosomal and intrachromosomal recombination and more recently from gene targeting experiments (13), that a significant variation in sequence homology between the two elements involved in the genetic exchange can reduce the homologous recombination/targeting frequency. The number and extent of polymorphic variation between two mouse strains in any given locus is unknown but may vary widely from gene to gene. In particular, introns are likely to be more divergent than exonic sequences. Although the importance of genetic drift on targeting frequency is currently undefined, to minimize the significance of this variation, the DNA used to construct the targeting vector should ideally be isolated from the same mouse strain as the cells used in the targeting experiments.

4.2 Enrichment schemes for targeted clones in culture

When a targeting vector is transfected into cells it may either integrate into its target locus or into a random chromosomal location. The relative ratio of these two pathways depends upon a number of factors which can not be experimentally controlled, such as the location of the target gene in the genome. In most cases, the frequency of random integration is far greater than for targeted recombination. While factors such as increasing the amount of sequence homology or using an insertion vector can increase the representation of targeted clones within a transfected population, it is also possible to significantly reduce the number of clones with random integration events in a population by requiring that the positive selection cassette should utilize elements of the target locus which are not present in the vector, for efficient transcription of the positive selection cassette. Vectors of this type have been described as enhancer, promoter, or polyadenylation trap vectors. These vectors are detailed in *Figure 6*. It is also possible to include negative selection cassettes which will be removed during a simple gene replacement event, since such markers are often included in a random integration event these non-targeted clones can be selected against.

4.3 Positive selection for targeted clones: enhancer and promoter trap targeting vectors

Promoter trap targeting vectors are designed to use the transcriptional machinery of the endogenous target gene to drive the positive selection cassette cloned in the targeting vector (*Figure 6A*). Vectors of this type are thus restricted to genes which are transcriptionally active in the target cells. Typically, the positive selection cassette is either cloned in-frame with the endogenous translated product, or the cassette may encode its own translation initiation site and is positioned upstream or in place of the normal translational start site. In either case the homologous sequences used in the vector are truncated so that they do not retain any promoter activity, thus, the majority of transfected clones in which the vector has integrated at random will be transcriptionally silent and will not survive positive selection. Typically, promoter trap selection will yield an enrichment of about 100-fold for targeted clones and should work with both replacement or insertion vectors (15).

There is some risk with this strategy in that a fusion gene with the positive selection marker may not be functional, thus it is important to test for positive selection with a control vector (15). Good candidate genes for this selection scheme are those with high expression levels in the target cells, although genes with low expression levels have been successfully targeted with promoter trap vectors (16).

An enhancer trap targeting vector is conceptually similar to a promoter trap targeting vector. Such vectors use a selection marker with a weak position dependent promoter which can be activated if integration occurs in the proximity of a transcriptional enhancer element. The enrichment factors attained with such vectors will be less significant than with promoter traps, although vector design will be simpler because a fusion transcript/gene product is not required.

4.4 Positive selection for targeted clones: polyadenylation trap targeting vectors

Polyadenylation trap targeting vectors are designed to use the transcription termination/polyadenylation signals of the target gene to generate a stable hybrid transcript consisting of elements from both the target gene and the positive selection cassette (*Figure 6B*). To make a targeting vector which relies on polyadenylation selection, the positive selection cassette would have its own promoter. In contrast to promoter trap selection schemes, this type of positive selection should work for most genes irrespective of whether or not

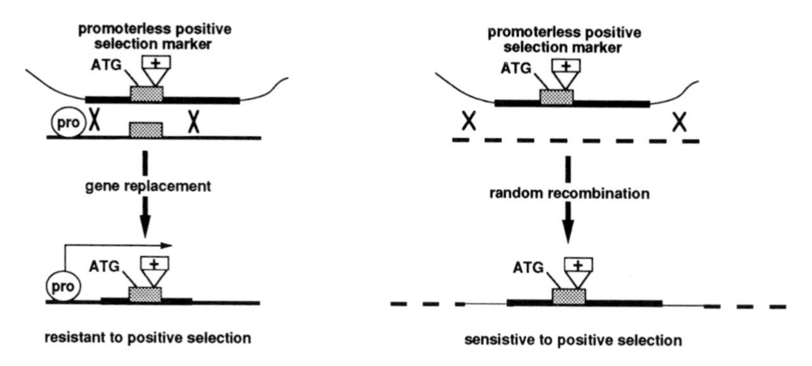

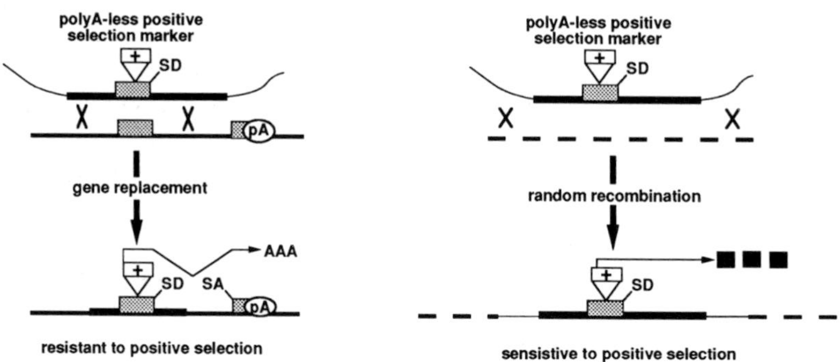

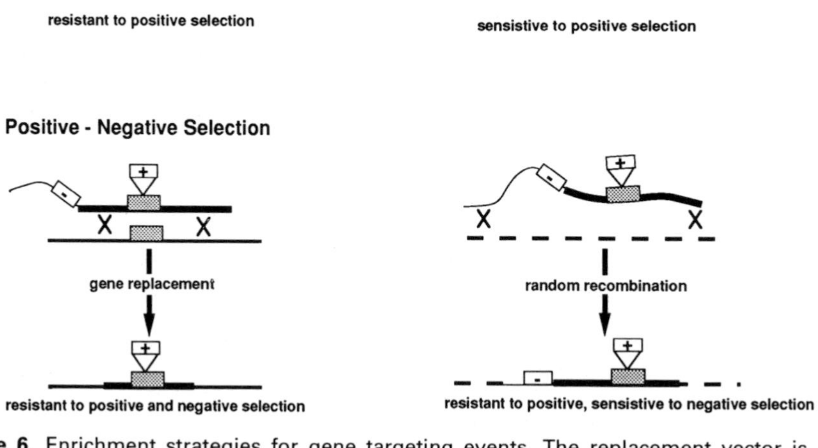

Figure 6. Enrichment strategies for gene targeting events. The replacement vector is linearized outside the target homology prior to transfection. The lines and boxes are as indicated in *Figure 2*. The dashed line is non-homologous chromosomal DNA. Strategies (A) and (B) may be used for replacement and insertion vectors. Strategy (C) is used for only replacement vectors. (A) Promoter trap. The translation start site is shown as ATG in the exon and the positive selection marker is fused in-frame. Alternatively, the positive

they are expressed in the transfected cells. The best location for the positive selection cassette is in an exon of the target gene so that the splice donor sequence of the disrupted exon can allow normal splicing to occur downstream of the expression cassette. The homologous sequences in the vector should be truncated to remove the endogenous transcription termination/polyadenylation signals. Typical enrichment factors obtained with such vectors have been in the range of 5–50-fold (for example, 17). In principle this selection scheme should work with both insertion and replacement vectors although the hybrid mRNA species generated with some fusions may be unstable. Consequently it is worthwhile to test the selection with an appropriate vector before initiating the gene targeting experiments.

4.5 Positive–negative selection for targeted clones

The techniques described so far have relied on the predicted structures of the target locus to select for targeted clones. It is also possible to use the components of the vector when integrated into random sites in the genome to select against non-targeted clones. This technique is known as positive–negative selection (18) and may be used with standard replacement vectors (*Figure 6C*). In this selection scheme the positive selection works exactly as previously described to isolate all of the clones which stably incorporate the vector DNA, irrespective of the integration site (targeted or random). The vector also contains a negatively selectable gene at one or both ends of the vector, during targeted gene replacement events this cassette is lost and degraded whereas clones in which the vector has integrated at random will typically incorporate this cassette. Thus, selection against this cassette will kill most clones of cells which have integrated the vector at a random location, while targeted clones will survive. Those clones which have integrated all of the vector components into the target locus by homologous recombination will also die under negative selection because of the inclusion of the negatively

---

selection marker may be cloned upstream of the translation start site or replace the translation start site of the target gene, provided the marker has a good consensus ATG for translation initiation, CC(A/G)CCAUGG (33). The promoter in the target locus is represented by a circled pro, and is not included in the target vector. Transcripts initiated by the endogenous promoter are represented by the horizontal thin arrow. Targeted clones use the endogenous promoter to generate an exon–positive selection marker fusion transcript; while, random recombinants are not usually adjacent to a promoter and not transcriptionally active. (B) Polyadenylation trap. The polyadenylation site, circled pA, is not included in the positive selection marker nor in the target homology. The splice donor site is represented by SD and splice acceptor by SA. The transcript initiated from the promoter of the positive selection marker is represented by a thin arrow. AAA, polyadenylation of the transcript. ■■■, transcript without polyadenylation. (C) Positive–negative selection. Both targeted and random integration events are selected with the positive selection marker. A targeted integration event removes the negative selection marker while the random recombination event integrates the negative selection marker; these clones survive and die in negative selection, respectively.

selectable gene. The enrichment achieved with negative selection is highly variable, the modal range in the literature is between two and 20-fold. Since the non-targeted clones which survive selection have usually mutated the negative selection cassette, it is possible to increase the efficiency by cloning an additional negative selection marker into the targeting vector. One virtue of positive–negative selection is that it does not depend upon elements of the targeted locus for its application. It is thus generally applicable to most replacement vectors, at any genetic locus irrespective of the structure or expression pattern. In contrast to the predictable structure of a targeted allele, the structure of the targeting vector when integrated at random sites in the genome is highly variable. Thus, the negative selection against such sites will be much less impressive than can be achieved with positive selection for promoterless or polyadenylation-less vectors.

5. Selection markers

As described previously, positive selection markers are a necessary component of a targeting vector because they facilitate the isolation of rare transfected cells from the majority of treated cells in a population. Negative selection markers are also a key component in that they facilitate the elimination of various subpopulations of transfected cells. A variety of selection markers have been described which act in either a dominant or recessive context. The dominant markers are more commonly used with ES cells since there are very few mutations that have been identified and are used for selection purposes, the one exception is the X-linked *hprt* gene (19). The selection cassettes summarized below are mainly derived from bacterial or viral origins, thus they are generally used as short intronless transcription units cloned adjacent to mammalian promoter/enhancer sequences and polyadenylation elements. A list of the commonly used cassettes and selections is detailed in *Table 1*.

5.1 Promoters and polyadenylation sites used for selection markers

The uses of appropriate promoter and transcription processing signals is very important to obtain adequate expression of the resistance markers in the targeting construct (*Table 2*). For the positive selection marker it is imperative that expression occurs at the target locus; for the negative selection marker expression should occur at all of the integration sites (random and targeted) generated in a transfection experiment. Thus, for most purposes, it is desirable to use promoters which express in the greatest number of genomic locations. The following promoters have been identified which appear to be relatively position independent in ES cells (6): phosphoglycerate kinase I (PGK-1), RNA polymerase II (pol 2), and β-actin. A number of position

Table 1. Commonly used selection cassettes

Dominant selectable markers

Selection cassette		Origin	Positive selection drug	Negative selection drug
Neomycin phosphotransferase	(*neo*)	B	G418	
Hygromycin B phosphotransferase	(*hph*)	B	Hygromycin B	
Xanthine/guanine phosphoribosyl transferase	(*gpt*)	B	Mycophenolic acid + HAT + Xanthine	6-thioxanthine
Herpes simplex thymidine kinase	(*HSVtk*)	V	—	GANC,FIAU
Diphtheria toxin	(*DT*)	B	—	none required

Recessive selectable markers

Selection cassette		Origin	Positive selection drug	Host genotype	Negative selection drug	Host genotype
Hypoxanthine phosphoribosyl transferase	(*hprt**)	M	HAT	*hprt*$^-$	6TG	*hprt*$^-$
Xanthine/guanine phosphoribosyl transferase	(*gpt*)	B	HAT	*hprt*$^-$	6TG 6TX	*hprt*$^-$ wt or *hprt*$^-$
Thymidine kinase	(*tk*)	M	HAT	*TK*$^-$	5BdU	*TK*$^-$
Herpes simplex thymidine kinase	(*HSVtk*)	V	HAT	*TK*$^-$	GANC,FIAU	wt

B = bacterial, V = viral, M = mammalian, HAT = amniopterin, hypoxanthine, thymidine, GANC = gancyclovir, FIAU = 1(1-2-deoxy-2-fluoro-β-Darabinofuransyl)-5-iodouracil, 6TG = 6-thioguanine, 6TX = 6-thioxanthine, 5BdU = 5-bromodeoxyuridine, wt = wild-type, * this is a minigene (19). 'Genotype' refers to the mutant background of the cell in which selection can be performed.

Table 2. Transfection efficiency of *neo* expression cassettes in ES cells (6).

	G418 ^{R}colonies 10^7 cells	Relative efficiency
pMC1neopA	13	1
RV4.0	1632	125
MC1neobpA	464	36
TKneobpA	124	10
Pol2sneobpA	324	25
Pol2neobpA	788	61
PGKneobpA	940	72

dependent weak promoters have been successfully used in gene targeting experiments, such as the synthetic mutant polyoma enhanced *HSVtk1* promoter, MC1 (9). In some contexts weak promoters can offer a selective advantage in the isolation of targeted clones provided there is adequate expression of the selection cassette at the target locus to give G418 resistance. This type of enrichment is like an enhancer trap. For negative selection purposes the MC1 promoter (9) appears to be adequate, this may be a function of the minimal expression levels required for toxicity and/or the possibility that the MC1 promoter may be up-regulated by the promoter/enhancers associated with the positive selectable markers present in the targeting vector.

The quality of RNA processing signals can also effect the efficiency of each selectable marker (6). For most purposes the PGK and bovine growth hormone polyadenylation sites have been shown to function efficiently.

6. Generating subtle mutations with gene targeting techniques

The strategies described in the previous sections have been used to establish disrupted alleles of a variety of genes in tissue culture cells and the mouse germline. These methods of gene inactivation have been widely used, but they require the introduction of a positive selection marker in order to select for the rare transfected cell, including targeted integrations. The positive selection marker also serves as the mutagen if inserted into the coding sequence since it generates a major disruption of the gene and introduces transcriptional control elements into the gene which may alter the expression of both the targeted locus and closely linked genes, this in turn may lead to difficulties in the interpretation of the phenotype with respect to genotype. Although major disruptions of a locus represent a valuable starting point for genetic analysis, a series of changes at the nucleotide level in both coding and control regions of the gene are important for the full understanding of the genes' function. Three techniques to introduce a small mutation into a gene are described which have been designed to avoid including a selectable marker in the recombinant locus.

Small mutations can be introduced into the genome in a single step, without including a positive selection marker, for instance with a replacement vector. However, if such a vector is used selection for cells that have taken up DNA would not be possible. This would make the identification of targeted clones very difficult since non-transfected cells would also be present and represent the vast majority of the population to be screened. Therefore, the vectors must be introduced into the cells under conditions where the transfected cells can be identified. In practice this requirement may be met by an efficient DNA delivery system such as microinjection (where each injected cell may be identified by location), or the co-introduction of a positive selection marker with the replacement vector where transfected cells can be selected. Alternatively, two step procedures have been devised to create subtle mutations.

6.1 Subtle mutations generated by microinjection

The introduction of DNA into cells by direct microinjection is very efficient, up to 20% of cells may integrate the injected DNA. Thus, if each injected cell is identified, non-selectable genetic modifications may be efficiently introduced into their target locus by homologous recombination (20). Thus after microinjection of the vector into the ES cells, they should be clonally expanded so that gene replacement events may be directly detected by Southern blot or PCR analysis. Although successful targeting has been reported in both fibroblast (21, 22) and ES (20) cells with microinjection there are severe technical constraints to this method, particularly when applied to ES cells. As a consequence, this technique is not widely practiced for ES cells. Readers interested in the practical application of this method should consult a specialist description of microinjection techniques (23).

6.2 Non-selectable mutations generated by co-electroporation

The co-introduction of two DNA molecules into a cell, where one is a positive selection marker and the other is a non-selectable vector (24, 25), can efficiently identify cells that have taken up DNA by positive selection. In fact, under a variety of different transfection protocols, the co-introduction of DNA is highly efficient (up to 75% of the cells which integrate DNA will include at least two copies). Although the majority of the transfected cells will contain concatemers of the targeting vector and the positive selection marker integrated into a single locus, 15% of the clones will have integrated the transfected DNA into different chromosomal sites. Thus, the co-introduction of two DNA molecules with a selection for one component and a screen of the selected clones for a targeted recombination event will generate three categories of clones; non-targeted clones, clones which have integrated concatemers of the targeting vector and the selection marker into the target locus, and clones targeted by simple gene replacement in which the positive selection marker has integrated in another location. This latter class of clone is the desired recombinant product.

6.2.1 Design of a co-electroporation vector

Co-electroporation vectors are configured as replacement vectors and they should be designed so that the recombinant clones can be readily identified by PCR and/or Southern blot analysis (see Section 2). In situations where the desired genetic modification would be minor, such as a single amino acid change, the ability to screen for targeted clones by either PCR or Southern blot analysis is best achieved by changing the wobble base pairs of a number of codons to create a unique PCR primer site or a novel restriction enzyme site. The protocol for co-electroporation is similar to that described in Chapter 2 (*Protocol 8*) for standard electroporation. However, the targeting vector and a plasmid containing a selectable marker are linearized and mixed together at a one to one molar ratio in PBS (Chapter 2, Section 2). A total of 25 to 100 micrograms (8.6 to 34 picomoles) of DNA should be used for each cuvette of cells in the electroporation. As a control for DNA toxicity, a cuvette of cells should be electroporated with the same amount of selectable marker DNA without the targeting vector.

6.2.2 Screening and analysis of co-electroporated targeted clones

The desired clone should have a recombinant allele with the designed mutation and the positive selection marker should be integrated elsewhere in the genome. Since the targeted allele should be the product of a simple gene replacement event many of the considerations for analysis have been previously described in Section 2.3.

Analysis of transfected clones should be done in two steps. First the clones should be screened for the mutation at the target locus by PCR and/or Southern blot analysis. Putative positive clones should be further analysed to ensure concatemers did not integrate into the target locus. The simplest diagnostic procedure for the absence of concatemers is to use a restriction enzyme that does not cut in either the target vector, the positive selection marker, or the bacterial plasmid, and hybridize the DNA with an external probe. If a concatemer has integrated into the target locus a very large fragment will be observed; whereas the target locus in the absence of concatemers will be of a predicted size. If a unique restriction site has not been identified, extensive digests are required which are hybridized to internal and external probes (described in Section 2.3).

6.3 Subtle mutations generated with a Hit and Run vector

A two step recombination technique is currently the best described and most widely used method for generating non-selectable mutations by gene targeting (*Figure 7A*). This technique relies on the provision of a positive selection marker to select for the primary targeted clones. This marker and other components of the vector are subsequently removed from the target locus by a second selectable intrachromosomal recombination step to generate the desired modified locus. This technique has been termed the 'Hit and Run' (26) or the 'In and Out' (27) targeting procedure. The first step utilizes an insertion vector where both the positive and negative selection markers are located outside the homologous sequences. The integration of the vector into the target gene will generate a duplication of genomic sequences as previously described (Section 3.2).

The second step relies on the spontaneous intrachromosomal recombination (pop-out) which will occur between the duplicate homologous sequences. This pop-out event results in the removal of the plasmid sequences, selection markers, and one complement of the duplication from the target locus. Since the negative selection marker is excised the cells revert and become resistant to drugs used against the negative selection cassette, consequently the revertant clones are readily selected from the population. The only limitation to this procedure is the spontaneous frequency of the loss of sensitivity to the negative selection agent due to events other than homologous recombination.

6.3.1 Design of a Hit and Run vector

Hit and Run vectors are modified insertion vectors which have a mutation in the homologous sequences. They also contain both a positive and a negative selection cassette in the vector backbone (*Figure 7B*). The targeted integration of an insertion vector has already been described (Section 3.2). However, an insertion vector which contains a mutation in the homologous sequences will form a variety of different integration products upon targeting. This is associated with gap formation and repair, mismatch heteroduplex repair, and the branch migration of the Holliday junctions which occurs during the targeted integration of an insertion vector. These factors all affect the position and presence of the mutation in the targeted allele (*Figure 7C*). For instance, the mutation may be removed and replaced with wild-type sequences by gap formation and repair, during integration into the target locus. The frequency with which the mutation is removed is related to the distance between the mutation and the double strand break; the closer the mutation is to the break the more frequently it is removed by gap enlargement and repaired using the chromosomal sequences as a template. A mutation which is located 100 base pairs from the break is frequently repaired (50–60% of the targeted clones), while a mutation located greater than one and a half kilobases from the gap is

A Hit and Run Pathway

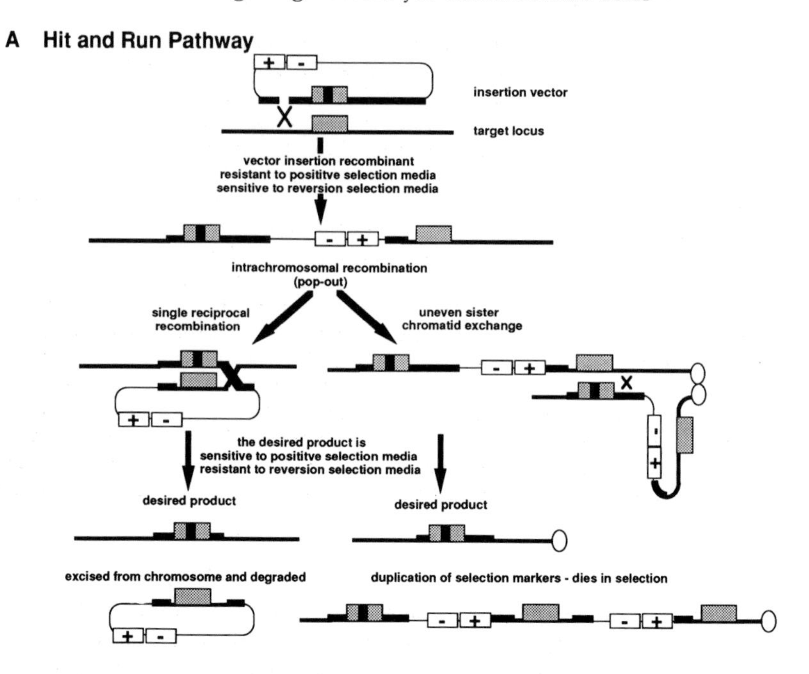

B Construction of a Hit and Run Vector

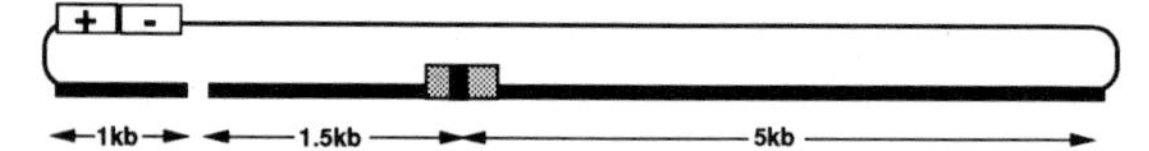

C Most Likely Integration Patterns

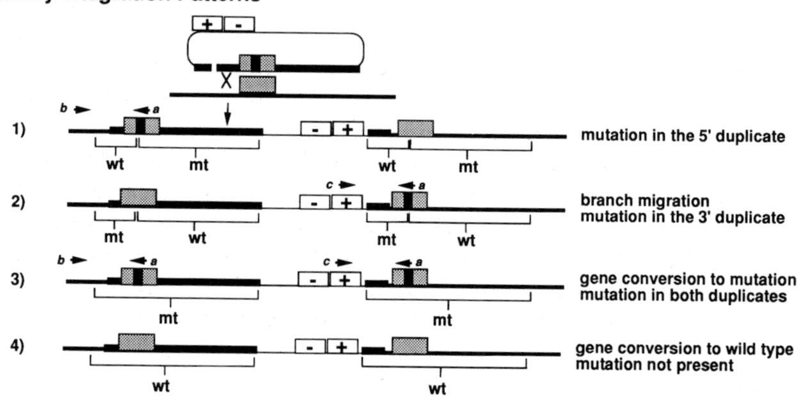

D Favorable Locations for the Mutation

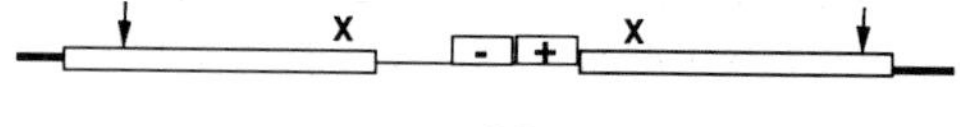

Figure 7. Hit and Run targeting. The insertion vector is linearized inside the target homology prior to transfection. The lines and boxes are as indicated in *Figure 2*. The small, non-selectable mutation is represented by a thick line in the exon. (A) The Hit and Run pathway. An insertion event followed by intrachromosomal recombination; either single reciprocal recombination (*left*) or uneven sister chromatid exchange (*right*). (B) The ideal design for a Hit and Run vector. The linearization site is on the short arm to encourage integration on the short arm and it is 1.5 kb away from the small mutation to decrease the likelihood that the mutation would be removed by gap formation and repair, or that the vector would integrate on the long arm due to branch migration of the Holliday junctions. (C) The most likely integration products with a Hit and Run vector. The brackets represent the duplicate homology available for the cross-over event which leaves the small mutation, mt, or restores the locus to wild-type, wt, after intrachromosomal recombination. The arrow heads represent PCR primers to specifically detect the mutation in both duplicates. Primer *a* is in the mutation, primer *b* is in the target locus outside the target homology, primer *c* is in the positive selection marker. (D) The most favourable positions for the mutation in the duplicate so that it remains after intrachromosomal recombination. The long rectangles represent the duplicate homologous sequences. The small arrows show the best locations for the mutation in the duplicate while the Xs show the worst.

---

infrequently repaired (5–10% of the targeted clones, P.H. and A.B. unpublished observations). The Holliday junctions (28) formed during the integration event often migrate. Thus, if the mutation is close to the double strand break (less than one and a half kilobases), the Holliday junctions will commonly migrate across the mutation and resolve on the other side of homologous sequences. This will result in the mutation appearing in the opposite duplicate than initially planned. Mismatch heteroduplex repair may result in the duplication or removal of the mutation in about 5–10% of the targeted clones.

The duplicated homologous sequences generated by a vector insertion event will recombine by intrachromosomal homologous recombination. This can occur either by single reciprocal recombination or more likely by uneven sister chromatid exchange (29). The frequency of these excision events is dependent on both the length and the degree of homology between the duplicates (30). To maximize the chance of recovering the mutation in the pop-out step the vector should be configured so that the mutation will appear at the distal position of either duplicate. To do this the mutation should be placed asymmetrically in the homologous sequences and the construct should be linearized on the short arm of homology. This maximizes the chance of integration by a cross-over on the short arm of the vector and positions the mutation at the distal end of one duplicate (*Figure 7B, C, D*). To avoid problems associated with gap formation and repair or the branch migration of Holliday junctions the mutation should be positioned at least one and a half kilobases from the linearization site.

A simple assay to identify the mutation is important. Ideally the mutation should be identifiable with a restriction enzyme site that gives a unique fragment on Southern blot analysis. In coding portions of a gene this may be achieved by altering the wobble nucleotides in the amino acid codons adjacent to the desired alteration. The mutation may also be designed to serve as a unique template for a PCR primer by using the wobble base pairs of codons to maximize the differences between the mutant and wild-type sequences.

6.3.2 Screening colonies for insertion and reversion events with Hit and Run vectors

The screen for vector insertion is done in two stages. First, the targeted clones are identified from the electroporated population which is resistant to the positive selection agent (see Section 3.2). After the clones with an insertion event have been isolated, it is necessary to verify that the mutation is present, both duplicates should be checked. If the mutation has a diagnostic restriction site then Southern blot analysis may be used. A probe internal or external to the construct will usually check both duplicates. If the mutation creates a unique primer site for PCR then try to develop two PCRs, one for each duplicate. One primer, specific for the mutation sequence, can be the same for both PCRs with each duplicate (*Figure 7C*). The second primer however is different for each duplicate and located immediately outside the target homology. The second primer for one duplicate should be in the chromosome and will only detect the mutation at that end of the duplicate (*Figure 7C*, 1 and 3); the second primer for the other duplicate should be located in the bacterial plasmid or in the selection markers and will only detect the mutation in the centre of the duplicate complex (*Figure 7C*, 2 and 3).

The same Southern blot analysis that identified the vector insertion event can be used to verify the pop-out event. The target locus in the pop-out clones should appear to be wild-type. If the revertant colonies appear wild-type then they should be screened for the presence of the mutation by Southern blot analysis or PCR.

6.3.3 Selection for Hit and Run targeting events

Hit and Run vectors contain both positive and negative selection cassettes to aid in the recovery of the insertion and pop-out events, respectively. The positive selection works in the manner described previously (Section 5). However, there are a number of important factors that should be taken into account prior to plating cells for reversion selection.

First, it is desirable to carefully choose the clones which are used for negative selection (*Figure 7C*). For instance, the clones which have the mutation in both duplicates, will recombine to leave the mutation in the chromosome irrespective of the position of the cross-over, thus these are the best candidates from which to recover the mutation. If such clones are not available, clones with the mutation in a single duplicate should be used, however, the chance of recovering a clone which has popped-out to the mutant or wild-type configuration depends on the position of the mutation within the duplicate (*Figure 7C* and *D*). Clones with the mutation located in the distal aspect of either duplicate will leave the mutation behind more frequently after the pop-out than those which have the mutation proximal to the vector backbone.

Secondly, it is important to be aware of the background problems which occur with negative selection. The background with reversion selection is typically much higher than with positive selection for a variety of reasons. When using negative selection it is very important to be certain that the cells were clonally derived. The propensity of ES cells to aggregate and cross-feed can enable wild-type cells to be carried as a minor fraction of a clone even under positive selection. Although minor contaminants may not always be detected by Southern blot analysis, if the background appears to be significant (10^{-3} to 10^{-4}) then clonality can usually be restored by single cell cloning or by successive low density plating under positive selection.

The extent of cross-feeding that occurs under negative selection is significant, particularly with ES cells. Previously, we have discussed negative selection in the context of positive–negative selection where the cells expressing the negative marker are greatly outnumbered by wild-type cells (approximately 10 000-fold). Consequently, the effects of cross-feeding that result in bystander killing is relatively minimal under these conditions. However, for reversion selection the vast majority of the cells express the negative marker, with just a few revertant cells being present in the population. Consequently the revertant cells will always have cells expressing the negative selectable marker as neighbours and thus are susceptible to bystander killing effects. With negative selection markers such as *hprt* and *gpt* it is necessary to recover *hprt*$^-$ ($6TG^R$) clones from predominantly *hprt*$^+$ populations with low plating densities (10^5 cells/90 mm plate), (31). Since the DT (*Table 1*) cassette is constitutively active it can not be used in a Hit and Run vector.

Since it is desirable to recover separate clonal reversion events, it is necessary to minimize the representation of clones which have popped-out or mutated the negatively selectable marker early in the clonal expansion of a line. This can be achieved by maintaining the clone under positive selection until the cells are plated under negative selection. This plating should be done under conditions where it is possible to establish both the background and the frequency with which new pop-out clones arise. The following steps should be taken to isolate clones with pop-out events during a Hit and Run targeting experiment.

(a) Expand the cells under positive selection (G418).
(b) Make eight 100 mm plates without G418, containing 1×10^5 targeted cells with feeder cells as described in Chapter 2.
(c) Add the negative selection drug at the appropriate concentration to one plate at the time of plating and label it Day 0; clones that grow on this plate represent the background.

(d) The next day, add the negative selection drug to a second plate and label it Day 1. Continue this routine for six more days to complete all eight plates.

(e) Resistant colonies should be conspicuous by ten days. Pick these clones and screen them for the pop-out event as described previously.

A sample of clones from all eight plates should be analysed to detect pop-out clones. While all of the plates will contain pop-out clones the relative ratio of these to the background will vary from one time point to the next.

7. Summary

In this chapter we hope to have conveyed the impression that the most important aspect of any gene targeting experiment is vector design since this will ultimately determine the progress of downstream experimentation, not only in the generation of the recombinant allele, but also in the analysis of the resultant cells or animal. However, at present it is not possible to predict what frequency of targeting will be obtained with any vector design or particular gene. This may vary over orders of magnitude. Clearly there is a rich diversity of experimental strategies which may be employed singularly or in combination as an aid in the identification of recombinant clones. We have detailed strategies which are designed to optimize the targeting frequency, and which make it possible to generate different mutations from gene ablation to the modification of a few nucleotides in a complex genome (32).

Acknowledgements

We would like to thank Richard Behringer for critical reading of this manuscript. P.H. acknowledges support from the Cystic Fibrosis Foundation. A.B. is a Leukaemia Society Scholar and the work described was supported by grants from the NIH, CF Foundation, and the Searle Scholars Program.

References

1. Lin, F.-L., Sperle, K., and Steernberg, N. (1985). *Proc. Natl Acad. Sci. U.S.A.*, **82,** 1391.
2. Smithies, O., Gregg, R. G., Boggs, S. S., Koralewski, M. A., and Kucherlapati, R. S. (1985). *Nature,* **317,** 230.
3. Robberson, B. L., Cote, G. J., and Berget, S. M. (1990). *Mol. Cell. Biol.,* **10,** 84.
4. Mombaerts, P., Clarke, A. R., Hooper, M. L., and Tonegawa, S. (1991). *Proc. Natl Acad. Sci. U.S.A.,* **88,** 3084.
5. Joyner, A. L., Skarnes, W. C., and Rossant, J. (1989). *Nature,* **338,** 153.
6. Soriano, P., Montgomery, C., Geske, R., and Bradley, A. (1991). *Cell,* **64,** 693.
7. Hasty, P., Rivera-Pérez, J., Chang, C., and Bradley, A. (1991). *Mol. Cell. Biol.,* **11,** 4509.

8. Ramírez-Solis, R., Rivera-Pérez, J., Wallace, J. D., Wims, M., Zheng, H., and Bradley, A. (1992). *Anal. Biochem.*, **201**, 331.
9. Thomas, K. R. and Capecchi, M. R. (1987). *Cell,* **51**, 503.
10. Moens, C. B., Auerbach, A. B., Conlon, R. A., Joyner, A. J., and Rossant, J. (1992). *Genes Dev.*, **6**, 691.
11. Jasin, M. and Berg, P. (1988). *Genes Dev.*, **2**, 1353.
12. Hasty, P., Rivera-Pérez, J., and Bradley, A. (1991). *Mol. Cell. Biol.*, **11**, 5586.
13. te Riele, H., Robanus Maandag, E., and Berns, A. (1992). *Proc. Natl Acad. Sci. U.S.A.*, **89**, 5128.
14. Thomas, K. R., Deng, C., and Capecchi, M. R. (1992). *Mol. Cell. Biol.*, **12**, 2919.
15. Schwartzberg, P. L., Robertson, E. J., and Goff, S. P. (1990). *Proc. Natl Acad. Sci. U.S.A.*, **87**, 3210.
16. Jeannotte, L., Ruiz, J. C., and Robertson, E. J. (1991). *Mol. Cell. Biol.*, **11**, 5578.
17. Donehower, L. A., Harvey, M., Slagle, B. L., McArthur, M. J., Montgomery, C. A., Butel, J. S., and Bradley, A. (1992). *Nature,* **356**, 215.
18. Mansour, S. L., Thomas, K. R., and Capecchi, M. R. (1988). *Nature,* **336**, 348.
19. Reid, L. H., Gregg, R. G., Smithies, O., and Koller, B. H. (1990). *Proc. Natl Acad. Sci. U.S.A.*, **87**, 4299.
20. Zimmer, A. and Gruss, P. (1989). *Nature,* **338**, 150.
21. Thomas, K. R. and Capecchi, M. R. (1986). *Cell,* **44**, 419.
22. Brinster, R. L., Braun, R. E., Lo, D., Avarbock, M. R., Oram, F., and Brinster, R. D. (1989). *Proc. Natl Acad. Sci. U.S.A.*, **86**, 7087.
23. Lovell-Badge, R. H. (1987). Introduction of DNA into embryonic stem cells. In *Teratocarcinomas and embryonic stem cells: a practical approach* (ed. E. J. Robertson), pp. 153–81. IRL Press, Oxford.
24. Davis, A., Wims, M., and Bradley, A. (1992). *Mol. Cell. Biol.*, **12**, 2769.
25. Reid, L., Shesley, E., Kim, H.-S., and Smithies, O. (1991). *Mol. Cell. Biol.*, **11**, 2769.
26. Hasty, P., Ramírez-Solis, R., Krumlauf, R., and Bradley, A. (1991). *Nature,* **350**, 243.
27. Valancius, V. and Smithies, O. (1991). *Mol. Cell. Biol.*, **11**, 1402.
28. Holliday, R. (1964). *Genet. Res.*, **5**, 282.
29. Bollag, R. J. and Liskay, M. (1991). *Mol. Cell. Biol.*, **11**, 4839.
30. Bradley, A., Ramírez-Solis, R., Zheng, H., Hasty, P., and Davis, A. (1992). In *Post implantation development of the mouse*, pp. 256–76 (Ciba Foundation Symposium 165). Wiley, Chichester.
31. Hooper, M. L. (1987). Isolation of genetic variants and fusion hybrids from embryonal carcinoma cell lines. In *Teratocarcinomas and embryonic stem cells: a practical approach* (ed. E. J. Robertson), pp. 51–70. IRL Press, Oxford.
32. Ramirez-Solis, R., Zheng, H., Whiting, J., Krumlauf, R., and Bradley, A. (1993). *Cell*, **73**, 279.
33. Kozak, M. (1984). *Nucleic Acids Res.*, **12**, 857.
34. Macejak, D. G. and Sarnow, P. (1991). *Nature,* **353**, 90.

2

Production of targeted embryonic stem cell clones

WOLFGANG WURST and ALEXANDRA L. JOYNER

1. Introduction

The discovery that cloned DNA introduced into tissue culture cells can undergo homologous recombination at specific chromosomal loci has revolutionized our ability to study gene function *in vitro* and *in vivo*. In theory this technique will allow us to generate any type of mutation in any cloned gene. The kinds of mutations that can be created include null mutations, point mutations, deletions of specific functional domains, exchanges of functional domains from related genes, and gain of function mutations in which exogenous cDNA sequences are inserted adjacent to endogenous regulatory sequences. In principle, such specific genetic alterations can be made in any cell line growing in culture. However, the limitation of this approach is that not all cell types can be maintained in tissue culture. Over ten years ago, pluripotent mouse embryonic stem (ES) cells derived from inner cell mass (ICM) cells of mouse blastocysts were isolated and conditions defined for maintaining them in culture (1, 2). ES cells resemble in many aspects ICM cells especially in their ability to contribute to all tissues in chimeras. Using stringent culture conditions, these cells can maintain their embryonic developmental potential even after many passages and following genetic manipulations. Genetic alterations introduced into ES cells in this way can be transmitted into the germline by producing ES cell chimeras (described in Chapters 4 and 5). Applying gene targeting technology to ES cells in culture therefore gives the opportunity to alter or modify endogenous genes and study their function *in vivo*. In the initial studies, one of the main challenges for such an approach was the identification of rare homologous targeting events amongst random integrations (discussed in Chapter 1). However, recent advances in selection schemes, in cell culture and freezing procedures, in vector construction using isogenic DNA, and in the development and application of the polymerase chain reaction has made it possible to identify these homologous recombination events efficiently.

The establishment of ES cell lines, conditions for their maintenance, and some manipulations have been covered in this series by Robertson (3). In this chapter we describe the basic techniques of ES cell maintenance, but concentrate on describing recent approaches for introducing DNA into cells for homologous recombination, as well as selection and screening procedures for identifying and recovering targeted cell clones. Although these techniques have been developed specifically for ES cells, they could easily be adapted for any tissue culture cell line.

2. Propagation and maintenance of embryonic stem cells

A number of different ES cell lines are available (see Chapter 4, *Table 1*) and it is advisable to use the recommended growth media and follow the suggested culture conditions for each particular cell line. We have used primarily the cell lines D3(4) and R1 (see Chapter 5), thus we describe methods for these cell lines.

2.1 General conditions

A well equipped tissue culture facility is essential and should include the following:

- laminar flow cabinet
- humidified incubator
- inverted microscope with a range of phase contrast objectives (×10 to ×25)
- binocular dissecting microscope with a transmitted light source
- liquid nitrogen storage tanks
- table-top centrifuge
- water-bath

All tissue culture procedures described must be carried out under sterile conditions using sterile plasticware and detergent-free glassware. Water quality is very important, therefore we recommend a Millipore Q filtration system (Millipore) for water purification or ultrapure water that can be purchased from commercial suppliers, (e.g. Gibco). Medium or solutions should be warmed to room temperature (or 37°C) before cells are placed in them.

2.2 ES cell culture media and solutions

For all tissue culture procedures we use Dulbecco's modified Eagle's medium (DMEM) with high glucose and glutamine, (e.g. Flow Labs, powder, Cat. No. 430-1600). The DMEM powder is dissolved in distilled water (Milli Q filtrate), buffered with bicarbonate, adjusted to pH 7.2 with HCl, and then filtered through a 0.22 μm nitrocellulose filter (S&S) into sterile bottles.

DMEM can also be purchased in liquid form and it can be stored for long periods (two to four months) at 4°C. It should, however, after 12 days of storage be supplemented with 2 mM glutamine (from a frozen 100 × stock at −20°C) because glutamine is unstable.

D3 ES cell growth media should contain the following:

- DMEM high glucose
- 0.1 mM non-essential amino acids (100 × stock, Gibco Cat. No. 320-1140AG)
- 1 mM sodium pyruvate (100 × stock, Gibco Cat. No. 320-1360AG)
- 10^{-6} M β-mercaptoethanol (100 × stock, stored at −20°C, Sigma Cat. No. 600564AG)
- 2 mM L-glutamine (100 × stock, Gibco Cat. No. 320-5030AG)
- 15% fetal calf serum (FCS)
- penicillin and streptomycin (final concentration 50 μg/ml each, Gibco Cat. No. 600-564AG)[a]

The same medium is used for R1 cells with the addition of 1000 U/ml LIF (Gibco) for growth on STO feeders.

The quality of FCS is very important for the maintenance of ES cells. We recommend that different batches be tested from different suppliers (see *Protocol 5*) for their ability to support growth of pluripotent ES cells. Suitable sera batches should then be ordered in large quantities and stored at −20°C for up to two years.

We use PBS (without Ca^{2+} and Mg^{2+}) for washing embryos prior to preparation of primary embryonic fibroblast cells and for rinsing ES cell layers before trypsinization.

1 litre of PBS solution made in H_2O contains the following:

- 10 g NaCl
- 0.25 g KCl
- 1.44 g Na_2HPO_4
- 0.25 g KH_2PO_4

The pH must be adjusted to 7.2 and the solution is then autoclaved.

For routine passaging of ES cells we disaggregate them by using 0.05% trypsin dissolved in Tris-saline/EDTA.

1 litre of 5 × trypsin/Tris-saline stock solution should contain the following:

- 8 g NaCl
- 0.4 g KCl
- 0.1 g Na_2HPO_4

[a] Can be added to the medium but it is not essential (see discussion Chapter 4, Section 2.1.1).

- 1.0 g glucose
- 3.0 g Trizmabase
- 2.5 g trypsin (Difco Cat. No. 0153-59)

The pH should be adjusted to 7.6, the solution sterilized through a 0.22 μm filter, and stored in 10 ml aliquots at −20°C.

This stock is diluted 1:4 with saline/EDTA before use as a 0.05% trypsin/ EDTA solution.

1 litre of saline/EDTA solution contains the following:

- 0.2 g EDTA
- 8.0 g NaCl
- 0.2 g KCl
- 1.15 g Na_2HPO_4
- 0.2 g KH_2PO_4
- 0.01 g phenol red
- 0.2 g glucose

The pH should be adjusted to pH = 7.2, the solution sterilized through a 0.22 μm filter, and stored at room temperature.

2.3 Production of fibroblast feeder layers

For long-term culture and maintenance, pluripotent ES cells must be grown either on monolayers of mitotically inactivated fibroblast cells (1, 4), or on gelatinized tissue culture dishes in Buffalo rat liver (BRL) cell conditioned medium (5), or in the presence of leukaemia inhibitory factor (LIF) (6, 7). Primary embryonic fibroblast (EMFI) cells (4) or the STO fibroblast cell line (1, 2, 8) are the most commonly used feeder layers. Most ES cell lines have been established on EMFI or STO cells and it is recommended to maintain each ES cell line on the feeder type on which it was originally established.

The D3 ES cell line was established on EMFI feeders and so we preferentially use these as feeder layers (4). The major disadvantage of EMFI cells is their short life span (15–20 cell divisions), which necessitates that new stocks of frozen vials of cells be made on a regular basis. We recommend the following protocol to establish a large stock of EMFIs. However, it is important to consider that each batch of EMFI cells that are prepared will have a different quality. Therefore each batch should be tested prior to usage for doubling time and the number of cell divisions before senescence. We have found no obvious difference between EMFI cells made from CD1 outbred, C57BL/6J inbred, or transgenic mice expressing *neo*.

Protocol 1. Preparation of a stock of EMFI cells

Materials

- 14–16 days post-coitus (dpc) pregnant mice
- sterile dissecting instruments (washed in 70% EtOH)
- screen, 1 mm diameter mesh (autoclaved)
- 50 ml glass beads (5 mm diameter) (autoclaved in a flask)
- stir bars (autoclaved)
- 500 ml Erlenmeyer flasks with foil or stoppers (autoclaved)
- plunger from a 5 ml disposable syringe
- 150 mm tissue culture dishes
- freezing vials, (e.g. Nunc Cat. No. 366 656)

Solutions

- PBS (without Ca and Mg) (Section 2.2)
- 0.05% trypsin/EDTA (Section 2.2)
- DMEM + 10% FCS
- DNase I (10 mg/ml)
- trypan blue (Flow Labs Cat. No. 16-910-49)
- 1 × freezing medium (DMEM, 25% FCS, 10% DMSO)

Method

1. Kill the pregnant mouse (about 15 dpc).
2. Moisten the belly with 70% ethanol and dissect out the uterus (see Chapter 4).
3. Place the uterus in a 10 mm Petri dish containing PBS, and dissect the embryos away from the uterus and all the membranes (see Chapter 6).
4. Transfer the embryos into a new dish containing PBS.
5. Remove heads and all internal organs (liver, heart, kidney, lung, and intestine).
6. Wash 8–10 carcasses in a 50 ml Falcon tube in 50 ml PBS, at least twice, to remove as much blood as possible.
7. Mince the carcasses in 2 ml PBS into cubes of about 2–3 mm in diameter with watchmaker's scissors in a 35 mm Petri dish.
8. Using the syringe plunger, press the cubes from the 10–12 embryos through the screen into an Erlenmeyer flask that contains 20 ml of glass beads and a stir bar.

Protocol 1. *Continued*

9. Flush 50 ml of trypsin/EDTA over the remaining clumps on the screen.
10. If the solution becomes too viscous due to released DNA, add 200 μl DNase I (10 mg/ml), and incubate the suspension at 37°C, for 30 min with stirring.
11. Add an additional 50 ml trypsin/EDTA and stir for another 30 min.
12. Repeat step 11.
13. Decant cell suspension and centrifuge cells (270 *g* for 5 min.).
14. Wash the cell pellet twice in DMEM + 10% FCS to eliminate trypsin activity, and resuspend the pellet in 10 ml DMEM + 10% FCS. If the cell pellet is too viscous add another 200 μl DNase I (10 mg/ml) and incubate for a further 30 min at 37°C.
15. Count viable nucleated cells using trypan blue. You should get about 5×10^7–10^8 viable cells from ten embryos.
16. Plate 5×10^6 nucleated cells per 150 mm tissue culture dish in 25 ml DMEM + 10% FCS, and culture overnight at 37°C (5% CO_2 in air).
17. Change the medium after 24 h to remove cell debris.
18. After two to three days of culture the EMFIs should form a confluent monolayer. Trypsinize each plate and replate on to five further plates (150 mm).
19. When the plates are confluent (usually after two or three days) freeze all the cells from each plate in one freezing vial in 1 ml of 1 × freezing medium and store at −70°C for one day.
20. Transfer the vials to liquid nitrogen.

Protocol 2. Preparation of EMFI feeder layers

Materials

- frozen vials of primary embryo fibroblasts (*Protocol 1*)
- tissue culture dishes
- PBS without Ca and Mg (Section 2.2)
- trypsin/EDTA (Section 2.2)
- DMEM + 10% FCS
- mitomycin C, (e.g. Sigma Cat. No. M-0503) (stock 1 mg/1 ml PBS stored in dark at 4°C and used within two weeks)

Method

1. Thaw a frozen vial of EMFI cells quickly at 37°C.
2. Add 10 ml DMEM + 10% FCS and centrifuge (270 *g*, 5 min).
3. Resuspend the cell pellet gently in 10 ml DMEM + 10% FCS and split on to five 150 mm plates each containing a total of 25 ml DMEM + 10% FCS.
4. Incubate cells at 37°C, 5% CO_2.
5. When the cells form a confluent monolayer (three days) they should either be trypsinized, split on to five 150 mm dishes and grown until they form a confluent monolayer (three days), or directly treated with mitomycin C.[a]
6. Remove the medium from the confluent plates and add 10 ml DMEM + 10% FCS containing 100 μl mitomycin C (1 mg/ml). Swirl plates to ensure an even distribution of medium.
7. Incubate cells at 37°C, 5% CO_2 for 2–2.5 h.
8. Wash the monolayer of cells twice with 10 ml PBS per dish.
9. Add 10 ml trypsin/EDTA to each plate.
10. Incubate 37°C, 5% CO_2 until the cells come off the plate (5–10 min).
11. Add 10 ml DMEM + 10% FCS to each plate and break any cell aggregates by gently pipetting.
12. Centrifuge cells (270 *g*, 5 min.) and resuspend the pellet in DMEM + 10% FCS.
13. Count the cells and dilute to a concentratin of 2×10^5 cells/ml.
14. Plate the cells immediately on to dishes suitable for ES cell culture containing DMEM +10% FCS. See *Table 1* for the appropriate cell densities for different plate sizes.
15. Allow feeders to attach overnight for best results or use them after 2 h.
16. Change the medium to ES cell medium (Section 2.2) before adding ES cells.

Mitomycin C treated EMFI feeders can be used for up to seven days with medium changes every three to four days.

[a] An alternative to steps 6–12 is to trypsinize the cells, rinse them, resuspend them in media, and treat them with 6000–10 000 rads of gamma irradiation.

The STO cell line is a thioguanine/oubain resistant subline of SIM mouse fibroblasts (9) that was established by Dr A. Bernstein (Mt Sinai Hospital, Toronto). This cell line was first found to support growth of pluripotent embryonal carcinoma (EC) cell lines (8, 10) and later to support growth of ES cell lines (1, 2). This cell line can be obtained from the American Type Culture Collection, but it is best to obtain a frozen aliquot from a laboratory using STO cells as a source of feeders for ES cells. This is because with many passages, the STO cell line changes and loses its ability to support optimal ES cell growth. For this reason, when you obtain a frozen vial of STO cells, it should be thawed, immediately expanded, and 20–40 vials of cells frozen away (2×10^6 cells/ml in 10% DMSO, 25% FCS). A new vial should then be thawed on to one 100 mm dish every four to six weeks. R1 cells were established on the SNL STO subline made by A. Bradley (11) although they can also be grown on EMFI feeders.

Protocol 3. Preparation of SNL STO feeder layers

Materials

- SNL STO cells
- tissue culture dishes
- PBS without Ca and Mg (Section 2.2)
- trypsin/EDTA (Section 2.2)
- DMEM + 10% FCS
- 0.1% gelatin, (e.g. Sigma or BDH) autoclaved in H_2O, and stored at 4°C.

Method

1. Prepare gelatinized plates as follows:
   (a) Cover the surface of the dishes completely with 0.1% gelatin solution (~ 5 ml for 10 cm plates).
   (b) Leave the dishes for approximately 2 min at room temperature.
   (c) Aspirate the gelatin solution and allow to dry before using the plates.
2. Thaw a vial of SNL STO cells and expand to the required number of 150 mm gelatinized dishes (see below).
3. Remove half of the medium (12.5 ml) from the confluent cultures and add 30 μl of mitomycin C (1 mg/ml) to each plate. Swirl each plate gently to mix the medium and to make sure the whole surface is covered.
4. Incubate the cells for about 2 h (37°C, 5% CO_2).
5. Follow steps 8–12 of *Protocol 2*.
6. Count the cells and dilute to a concentration of $2.5\text{–}3.5 \times 10^5$ cells/ml.
7. Plate the feeder cell suspension on to the size and number of gelatinized plates needed (see *Table 1* for recommended cell volumes).

We passage and prepare mitomycin C treated EMFI or STO feeder layers twice per week. In a typical schedule for EMFI cells, a vial of cells is thawed

Table 1. Volume of EMFI cells plated on different size tissue culture dishes

Plate diameter (cm)	Area (cm^2)	Relative size	ml medium per plate	Number of feeder plates from 1 × 14 cm plate
14	150	1	25	
9	63	1/2.4	10	3–4
6	20	1/7.5	5	6–8
3.5	9.6	1/15.6	2	18–20
1.5	1.8	1/80	0.5/well	80 wells

on Tuesday on to five 150 mm plates, one or two of these plates are passaged on Friday at a 1:5 dilution, and the medium is changed on the remaining plates. Feeder layers are made from the passaged cells and the remaining EMFIs passaged (1:5) on the following Monday. The passaged cells are then used on the next Thursday to prepare feeder layers. For SNL STO cells, we typically passage the cells on Mondays, plate them on to 150 mm dishes (8.75×10^6 cells/plate) for mitomycin C treatment on Thursday, and on to 100 mm dishes (3.5×10^6 cells/plate) for further expansion. On Thursday, the cells on the 100 mm dishes are passaged on to 150 mm dishes (8.75×10^6 cell/plate) for making feeder layers on Monday, and on to 100 mm dishes (3.5×10^6 cells/plate) for further passaging.

2.4 Growth of ES cells on feeder layers

In our experience each ES cell line has different requirements for continued growth in an undifferentiated state and maintenance of their ability to contribute to the germline. We have had most experience with D3 and R1 cells and the following protocol is recommended for these cell lines. For other cell lines, you should consult the laboratory where the cell line was generated. The parameters that need to be considered are:

- growth media
- length of growth between passages
- number of cells to be plated at each passage
- extent of trypsinization
- % CO_2 in the incubator
- feeder layer cell type

The main factor to consider is that no matter how stringent the culture conditions are, abnormal variant cell clones will arise in the ES cell population with time in culture. Some, but not all, of these variants will have an obvious abnormal karyotype and most are unlikely to be able to contribute to the germline. Some variants will have a growth advantage by having a shorter doubling time or better plating efficiency. Such abnormal variants can very quickly outgrow the normal cells and become the predominant cells in the culture. We therefore recommend that in general you should use a rich medium and passage the cells frequently (every two days) with a minimum dilution of the cells (1:5 to 1:7). If the cells require more frequent passage or a higher dilution, then an abnormal variant has probably taken over the culture. Karyotype analysis of ES cells (the parental line and subclones) can be used to monitor gross chromosomal changes but a seemingly normal karyotype does not guarantee germline transmission.

Another important factor to keep in mind is that all cells grown in culture can become infected with mycoplasma. Such infections do not always have an obvious effect on cell growth or morphology. However, mycoplasma can cause chromosome damage and reduce the efficiency of obtaining ES cell chimeras. Therefore, all ES cell lines and sublines as well as STO and EMFI feeder stocks should be routinely tested for mycoplasma after two weeks growth in the absence of antibiotics (see Chapter 4, Section 2.1.1 for further discussion).

Protocol 4. Growth of ES cells on feeders

Materials

- 60 mm dishes containing feeder layers (*Protocol 2* or *3*)
- ES cell medium (Section 2.2)
- trypsin/EDTA (Section 2.2)
- PBS without Ca and Mg (Section 2.2)

Method

1. Thaw quickly one vial of frozen ES cells (*Protocol 7*) and transfer cells to a 12 ml tube.
2. Add 10 ml ES cell medium and centrifuge 270 *g*, 5 min.
3. Resuspend in 5 ml ES cell medium and plate on a 60 mm dish of EMFIs.
4. Change the medium the next day.
5. On the second day (cells should be confluent, see *Figure 1A*) wash cells with PBS and add 2 ml trypsin/EDTA.
6. Incubate for 3–5 min or until cells begin to come off the plate (37°C, 5% CO_2).
7. Gently pipette the cells up and down to break cell clumps.
8. Add 5 ml ES cell medium and centrifuge (5 min., 270 *g*).
9. Aspirate supernatant and resuspend in 5–7 ml medium.

10. Pass 1 ml of the cell suspension (about 2×10^6 D3 cells or 1×10^6 R1 cells/60 mm dish) to a fresh 60 mm feeder layer containing 4 ml ES cell medium.
11. Change the medium the next day (if there are a lot of dead cells), and pass cells every second day as described above.

To test serum batches three concentrations of FCS (10%, 15%, 30%) for each test lot should be compared to a serum batch which is known to give good results (with respect to plating efficiency, colony morphology, and the toxicity to ES cell culture). The plating efficiency of ES cells at low density should be at least 15% and colonies should retain an undifferentiated morphology for a few days of growth once the colonies are visible under a microscope. The plating efficiency should be the same in the three concentrations of FCS, although growth is often fastest in 15% FCS.

Protocol 5. Screening batches of serum for optimal ES cell growth

Materials

- 35 mm dishes containing feeder layers (*Protocol 2* or *3*)
- ES medium (Section 2.2) with 10%, 15%, and 30% FCS
- FCS (control and test batches)[a]
- methylene blue (3.3 g methylene blue, 1.1 g basic fuchsin in 1 litre of methanol)
- PBS without Ca and Mg (Section 2.2)

Method

1. Plate 10^3 single ES cells on 35 mm dishes containing feeder layers in 10%, 15%, and 30% FCS medium. Prepare duplicates for each serum concentration.
2. Let the colonies grow and change medium every second day.
3. After seven to ten days of culture, rinse the plate with PBS.
4. Add 2 ml methylene blue to each plate for 5 min.
5. Rinse the plates several times with water.
6. Examine the plates with respect to the number of colonies, colony size, and their morphology. The morphology may be clearer before staining.

If one serum batch seems to be suitable, it can be further checked by growing ES cells in ES cell medium (15% FCS) for at least five passages (check the morphology and cell density after each passage).

[a] We have had good results with HyClone serum.

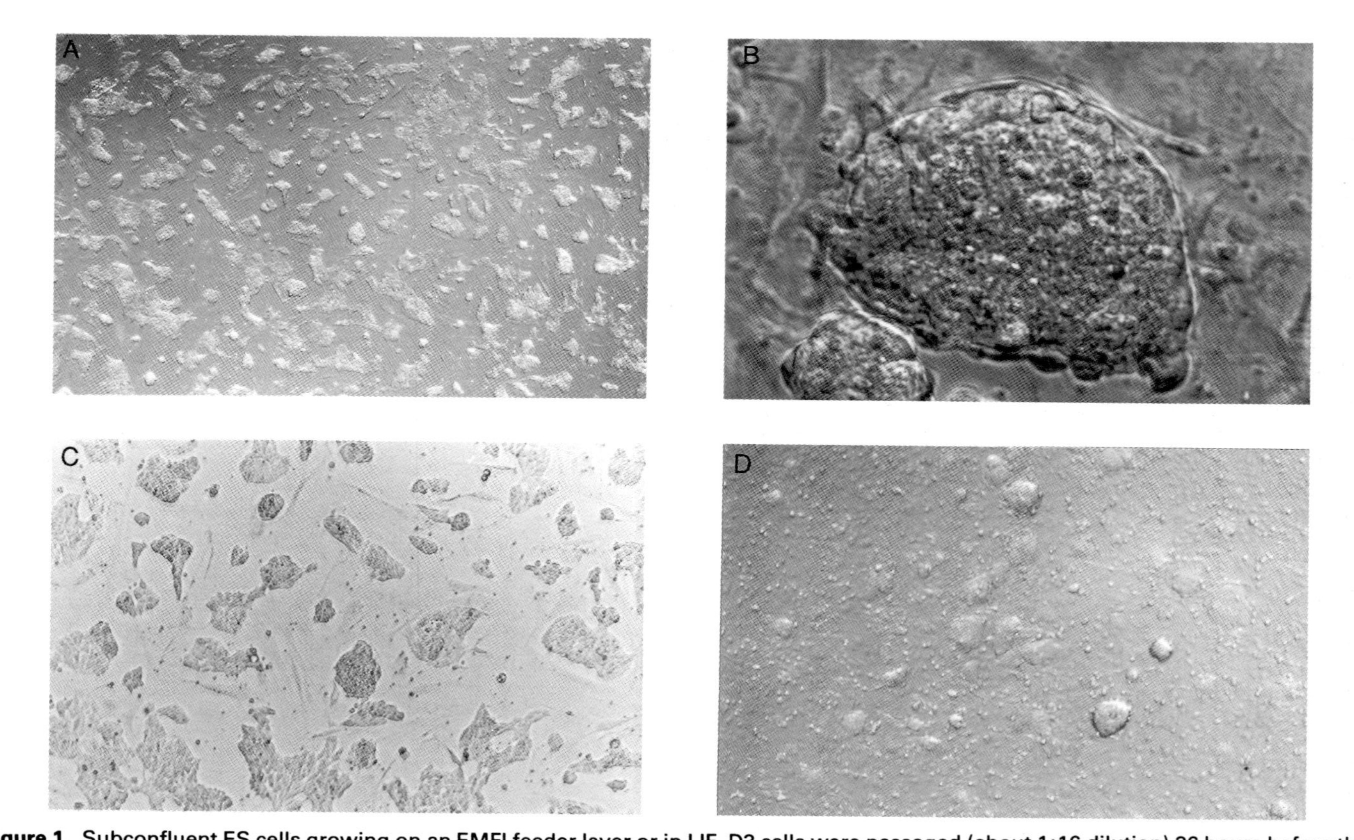

Figure 1. Subconfluent ES cells growing on an EMFI feeder layer or in LIF. D3 cells were passaged (about 1:16 dilution) 36 hours before the photograph was taken and plated on either (A, B) EMFI feeder layer or (C) a gelatinized plate in medium containing LIF. (D) R1 cells were similarly passaged on to SNL STO feeder cells. (A, D) Magnification 28 ×; (B) magnification 280 ×; (C) magnification 70 ×.

2.5 Growth of ES cells in leukaemia inhibitory factor (LIF)/differentiation inhibitory activity (DIA)

ES cells can be grown in an undifferentiated state without feeder cell layers on gelatinized plates (*Protocol 3*) in Buffalo rat liver (BRL) cell conditioned medium (5). The active factor in the conditioned medium that prevents ES cell differentiation and/or promotes undifferentiated cell proliferation is ES cell differentiation inhibitory activity (DIA) (5). Purified DIA is identical to the myeloid regulatory leukaemia inhibitory factor (LIF) (6, 7), which has been cloned and molecularly characterized (12, 13).

Supernatants from transfected Cos cells expressing recombinant LIF can have the same differentiation inhibitory activity as BRL conditioned medium (6). ES cells maintained without feeder layers in the presence of BRL cell conditioned medium or recombinant LIF are able to remain pluripotent since in chimeras they can contribute to all somatic tissues as well as to the germline (14). ES cell lines can also be established in LIF, however, the few lines that were established in LIF were found to have a high degree of chromosomal abnormalities (14). This suggests that ES cell lines established in LIF can not maintain their pluripotency when maintained for long periods in recombinant LIF. We therefore do not recommend extended growth of ES cells in LIF alone without feeders. It should be noted that ES cells grown in LIF have a more flattened morphology than cells grown on feeders (see *Figure 1C*). However, we have found that when the cells are returned to feeder layers they revert to their original, round morphology after one passage.

Recombinant LIF can either be purchased from commercial suppliers, (e.g. Gibco) or obtained from supernatants of Cos cells transfected with recombinant LIF constructs. If such supernatants are used, each batch should be carefully tested for its ability to maintain ES cells in an undifferentiated state as in *Protocol 5*. Commercial suppliers of recombinant LIF usually recommend a concentration of 1000 U/ml. However, each ES cell line might have different requirements and we suggest that it should be tested. We use 500 U/ml or 1000 U/ml LIF (Gibco) for short-term culture of D3 or R1 ES cells respectively.

Protocol 6. Propagation of ES cells in LIF

Materials

- gelatinized tissue culture plates (see *Protocol 3*)
- recombinant LIF (Gibco Cat. No. 3310SB)
- ES cell medium (Section 2.2)
- trypsin/EDTA (Section 2.2)
- PBS without Ca and Mg (Section 2.2)

Protocol 6. *Continued*

Method

1. On the second day after passage, wash cells once with PBS and add 3 ml trypsin/EDTA per 100 mm dish.
2. Incubate for about 3–5 min. (37°C, 5% CO_2) until the cells come off the plate.
3. Gently pipette up and down to break cell clumps.
4. Add 5 ml medium and centrifuge (5 min., 270 *g*).
5. Aspirate supernatant and resuspend the cell pellet in 5–7 ml ES cell medium.
6. Pass 1 ml (about 4×10^6 D3 or 2×10^6 R1 cells) to a gelatinized 100 mm plate containing 9 ml ES cell medium plus 500 or 1000 U/ml LIF.
7. Change medium next day and pass them every second day as above.

2.6 Freezing, storage, and thawing of ES cell lines

ES cells can be frozen like other tissue culture cells in 25% FCS using 10% DMSO as a cryoprotectant. As a general rule, freeze cells slowly and thaw them quickly. For long-term storage (indefinitely), cells should be kept under liquid nitrogen and for short-term storage (up to six months) they can be kept in a −70°C freezer. ES cells freeze best at a cell density of $5–10 \times 10^6$ cells per ml of 1 × freezing medium. This corresponds to approximately three tubes of 1 ml each per 100 mm dish. They can be frozen from different plate sizes using less freezing medium (60 mm dish 1 ml, 35 mm dish 0.5 ml, and 12 mm dish 0.25 ml). It is important to thaw the cells rapidly and remove the DMSO-containing medium as soon as possible. Also, the cells should be replated on the appropriate size plate (1 ml on to a 60 mm dish, 0.5 ml on to a 35 mm dish, and 0.25 ml on to a 12 mm dish).

Protocol 7. Long-term freezing of ES cell stocks

Materials

- freezing vials (1–2 ml, e.g. Nunc)
- styrofoam containers (Sarstedt)
- liquid nitrogen tank
- −70°C freezer
- ES cell 2 × freezing medium to be made up from fresh ES cell medium (final concentration: 50% FCS, 20% DMSO)
- ES cell medium (Section 2.2)

Method

1. Trypsinize cells (see *Protocol 4*) from a 100 mm dish (about 3 or 1.5×10^7 D3 or R1 cells respectively).
2. Pellet cells by centrifugation (270 *g*, 5 min.) and resuspend in 1.5 ml ES cell medium.
3. Aliquot 0.5 ml of the cell suspension into freezing vials on ice (4°C). Add 0.5 ml 2 × freezing medium to each tube.
4. Quickly shake the tube on ice and immediately transfer it to a pre-cooled styrofoam box in a −70°C freezer.
5. After 24 h transfer the tubes on dry-ice to liquid nitrogen.

3. Electroporation of ES cells

There are several different methods available to introduce DNA into embryonic stem cells, some of which have been described in this series by R. H. Lovell-Badge (15). Two approaches have been taken for gene targeting experiments; microinjection and electroporation. Microinjection of DNA into the nucleus of ES cells requires expensive equipment and a high degree of manual skills (15). The advantage of this route, however, is the high efficiency of DNA uptake (up to 20% of cells injected) which allows the creation of mutations without the use of selectable markers. To date, however, only one report has described the successful use of this sophisticated method (16). The more common and very successfully used approach is DNA transfer by electroporation. This has the advantage that it is technically relatively simple. The major disadvantage is the low transformation efficiency, which ranges from 10^{-3} to 10^{-4}, and thus necessitates the use of selectable marker genes in the targeting vectors (see Chapter 1 for details). However, advances in the development of selection schemes and the simplicity of the electroporation procedure make it the method of choice. Indeed, it has even been possible, using electroporation and recently developed two step selection schemes, to create subtle mutations in the absence of a marker gene in the endogenous locus (see Chapter 1, Section 6.3).

3.1 Standard electroporation conditions

DNA can be transfected into ES cells by application of a high voltage electrical pulse to a suspension of cells and DNA (17). After application of this pulse, the DNA passes through pores in the cell membrane. This procedure, however, results in the death of about 50% of the cells with conditions that give optimal transfection efficiency. The parameters which influence the efficiency and cell viability are the voltage, the ion concentration, the DNA concentration, and the cell concentration. Ideally, all these different parameters should be determined considering cell viability and transfection efficiency. The following protocol is based on our experience with the ES cell lines D3 and R1, and the fibroblast cell line rat 2.

Protocol 8. Standard electroporation of ES cells

Solutions

- ES cell medium (Section 2.2) supplemented with 500 or 1000 U/ml LIF (Gibco)
- PBS without Mg and Ca (Section 2.2)
- 100 mm gelatinized plates (*Protocol 3*)
- geneticin (G418, e.g. Gibco Cat. No. 860-1811IJ)
- gancyclovir (Cytovene, Syntex)

Materials and equipment

- electroporation apparatus (BioRad, gene pulser)
- electroporation cuvettes (BioRad, Cat. No. 165-2088)

Method

1. On the second day (36 h) after passaging recently thawed ES cells, trypsinize them to single cells (*Protocol 4*).
2. After pipetting the cells gently in 3 ml trypsin (for a 100 mm plate) to break up the clumps, add 7 ml medium and pipette gently up and down again.
3. Pre-plate the cells by incubating them in the dish for 30 or 15 min. (37°C, 5% CO_2) for D3 or R1 cells respectively to allow feeder cells to attach to the plate.
4. Harvest the ES cells carefully and transfer them to a 50 ml Falcon tube, combining the cells from two to five plates.
5. Centrifuge the cells for 5 min. at 270 *g*.
6. Aspirate the supernatant and resuspend the cells in cold PBS (1 ml/100 mm plate starting culture).
7. Count the ES cells and adjust cell concentration to 7×10^6 cells/ml (there should be approximately 2×10^7 D3 cells/100 mm dish).
8. Mix 0.8 ml of the cell suspension with 40 μg of linearized vector DNA[a] (for an approximately 10 kb vector), and transfer it into an electroporation cuvette.
9. Set up the electroporation conditions in advance (240 V, 500 μF for the BioRad gene pulser).
10. Transfer each cuvette into the cuvette holder with electrodes facing the output leads.
11. Deliver the electric pulse.
12. Remove each cuvette from the cuvette holder and place it on ice for 20 min.

13. Transfer the cell suspension from each cuvette into 10 or 20 ml of ES cell medium containing LIF.
14. Aliquot each 10 ml of the cell suspension into a 100 mm gelatinized dish.
15. The next day change the medium (supplemented with LIF).
16. Two days after the electroporation, add (in addition to LIF) the drugs for the selection (active G418 at 150–250 μg/ml, and gancyclovir at 2 μM; see also Sections 3.2 and 3.3).
17. Change the medium every day if working with GANC, otherwise every two days when there is little cell death.
18. After about eight days of selection, drug resistant colonies should have appeared (see *Figure 2*) and be ready to screen either by PCR or by Southern blot analysis (see Sections 4.1 and 4.2).

[a] The large scale vector DNA preparations should be purified over CsCl gradients. After restriction enzyme digestion the DNA should be extracted with phenol/chloroform, ethanol precipitated, and dissolved in sterile PBS (1 μg/μl) in a sterile hood.

3.2 Choice of antibiotic resistance genes for gene targeting vectors

Two of the most commonly used genes which confer antibiotic resistance in mammalian cells are the bacterial genes *neomycin* (*neo*; aminoglycoside phosphotransferase) and *hygromycin B phosphotransferase* (*hph*). Expression of these genes can be selected for with geneticin (G418) or hygromycin B respectively.

As discussed in Chapter 1 (Section 5.1) the most commonly used promoters for driving the expression of selectable genes in ES cells are the phosphoglycerate kinase (PGK-1), the herpes simplex thymidine kinase including polyoma enhancer fragments (pMC1), the RNA polymerase II, and the human β-actin promoter. We have found that in addition to the promoter, the sequence context of the translation start site (ATG) has a great influence on the percentage of cells that become drug resistant. With our β-actin–*neo* construct that contains 500 base pairs of the human promoter (18) and an artificial translation start that conforms to the Kozak consensus, we routinely obtain about 2500 neo^R colonies for 5.6×10^6 electroporated cells. With the bacterial *neo* translation start sequences we obtain ten-fold fewer colonies. For genes not expressed in ES cells, we recommend using the strongest of these vectors since they are subject to minimal position effects. However, this will also increase the number of neo^R colonies which have to be screened for homologous targeting events.

We routinely test the activity of different G418 batches for killing non-resistant ES cells in culture by setting up kill-curve experiments. We then use the lowest concentration of G418 that gives 100% killing within five days.

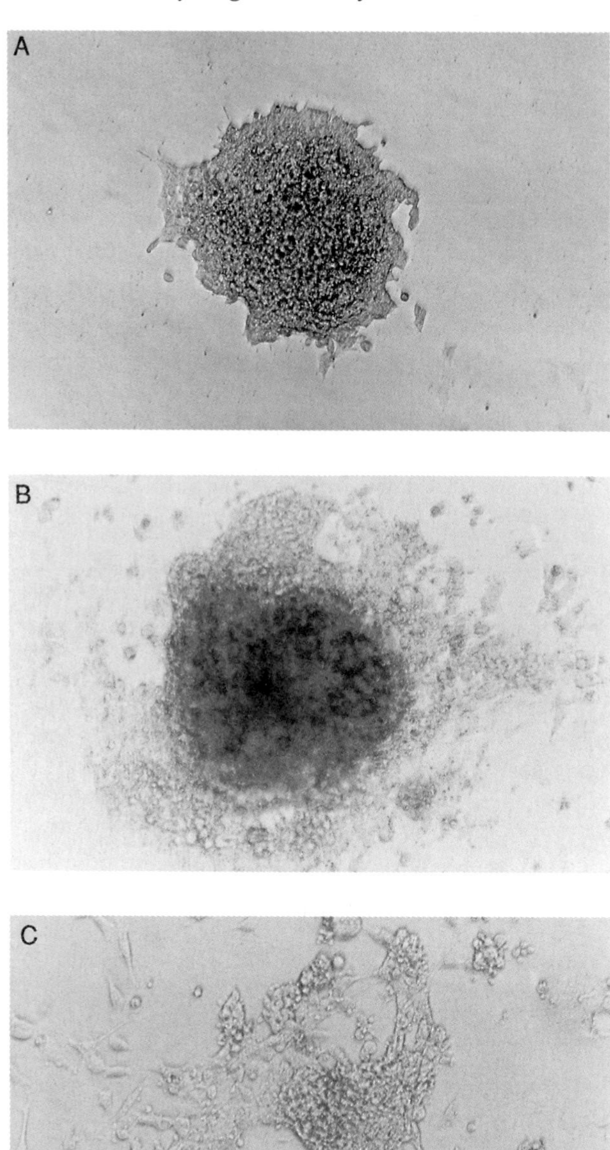

Figure 2. Cell morphology of ES cell colonies growing on a gelatinized plate in LIF. R1 cells were electroporated as in *Protocol 8* and grown in selective medium for eight days. Colonies have a number of different morphologies due to differentiation.

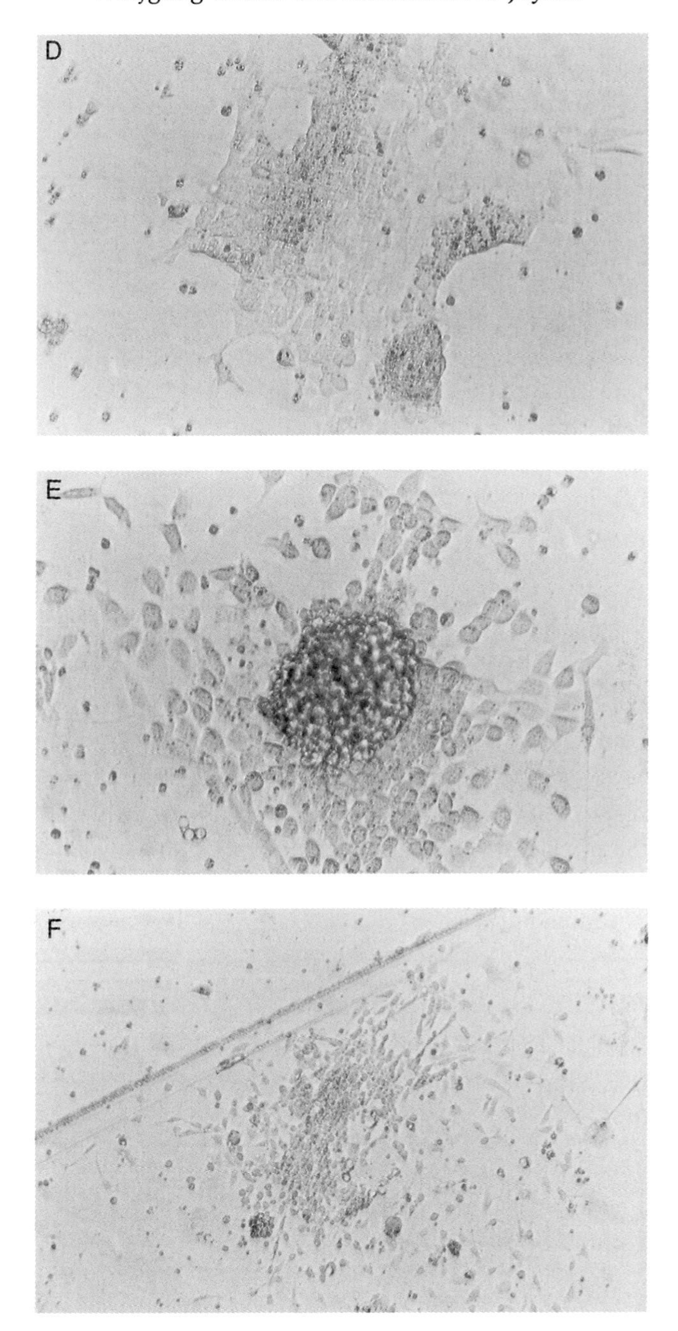

(A) Undifferentiated ideal ES cell colony. (B–F) Various differentiated colonies. D3 differentiated colonies look similar to those shown in (B, D). Magnification 75 ×.

The concentration we use varies between 150 and 250 μg/ml active G418 depending on the batch used. The G418 concentration needed will be higher if the ES cells are grown on *neo*^{R} EMFI or STO feeder layers. You will notice that the ES cells die faster in higher concentrations of G418.

Protocol 9. Antibiotic resistance test

Solutions

- ES cell medium (Section 2.2) (supplemented with LIF)
- geneticin, (e.g. Gibco Cat. No. 860-1811IJ; frozen stock: 40 mg active G418/ml in PBS)
- gelatinized plates (*Protocol 3*)
- methylene blue stain (3.3 g methylene blue, 1.1 g basic fuchsin in 1 litre of methanol)
- PBS without Ca and Mg (Section 2.2)

Method

1. Prepare ES cell medium containing 0, 50, 100, 150, 200, 250, 300 μg/ml of G418.
2. Plate 3–4 × 10^4 ES cells on to 60 mm gelatinized tissue culture dishes in ES cell medium containing LIF and each of the concentrations of G418, or similarly on to feeder plates without LIF.
3. Change the medium every second day for at least six days.
4. Wash the plates with PBS, stain with methylene blue solution for 5 min., and then wash several times with water.
5. Count the stained colonies.

---

4. Screening single colonies for homologous targeting events

One of the original limitations of gene targeting by homologous recombination was the difficulty in identifying cells containing the predicted modification at the targeted locus. However, the development and application of the polymerase chain reaction has allowed the rapid identification of rare homologous integration events. Progress also has been made in selection schemes, in the ES cell culture procedures, and in DNA isolation methods which have made it easier to screen large numbers of ES cell colonies for targeting events by Southern blot analysis.

Using electroporation as the DNA transfer technique, schemes have been developed to enrich for the rare homologous recombination events among the majority of random integrations (discussed in 19). Targeted mutations into selectable genes such as *hprt* are straightforward to identify using direct selection of *hprt*$^-$ cells. However, as discussed in Chapter 1, for most loci a selectable marker gene must be introduced into the targeting vector to select for cells that have integrated the vector. Since the frequency of homologous recombination events relative to random integration is in the order of 10^{-1} to 10^{-4}, different enrichment strategies have been applied to reduce the number of antibiotic resistant colonies produced due to random integration. The two most commonly used schemes are:

(a) for genes which are expressed in ES cells, promoterless *neo* cassettes
(b) for genes not expressed in ES cells an enrichment scheme termed positive–negative selection (PNS).

These and other schemes are discussed in detail in Chapter 1. With these strategies, enrichments of homologous to random integration have been reported in the range of ten to 100-fold.

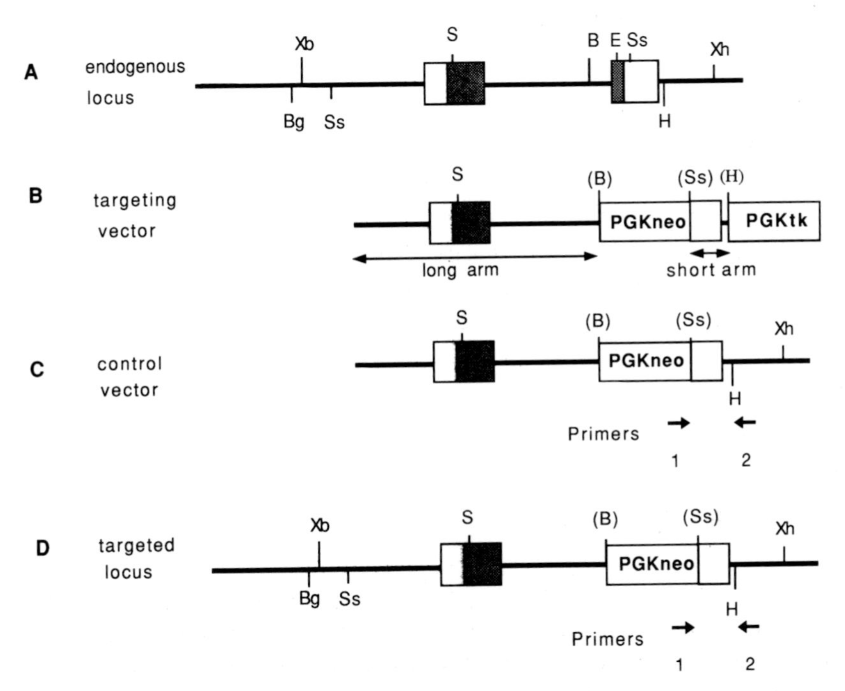

Figure 3. Construction of targeting and control vectors for screening by PCR. (A) An endogenous locus of a two exon gene. (B) A targeting vector with short and long arms of homology, a positive selection *neo* vector (PGK*neo*) replacing part of the second exon, and a *tk* vector (PGK*tk*) for negative selection at one end. (C) A control vector similar to the targeting vector but missing the *tk* vector and instead including a longer arm of homology. The primer binding sites for PCR are indicated as arrows. (D) The targeted locus following double cross-over homologous recombination. The exons are shown as rectangles with coding sequences indicated by shading. Bg, *Bgl*II; Xb, *Xba*I; Ss, *Sst*I; S, *Sal*I; B, *Bam*HI; E, *Eco*RI; *Hin*dIII; Xh, *Xho*I.

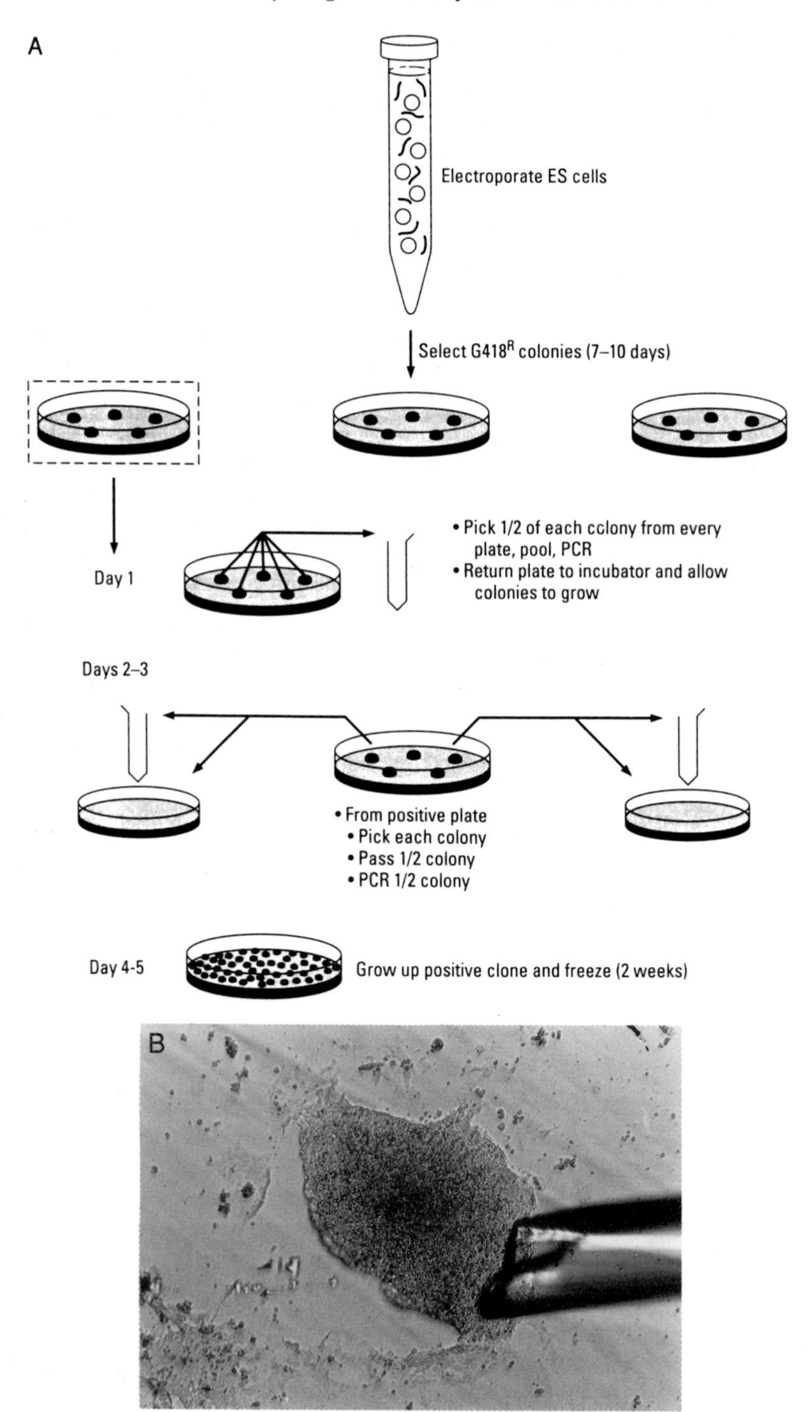
A
Electroporate ES cells
Select G418R colonies (7–10 days)
Day 1
• Pick 1/2 of each colony from every plate, pool, PCR
• Return plate to incubator and allow colonies to grow
Days 2–3
• From positive plate
• Pick each colony
• Pass 1/2 colony
• PCR 1/2 colony
Day 4-5
Grow up positive clone and freeze (2 weeks)
B

Figure 4. Screening for targeted colonies using PCR. (A) A schematic diagram showing the steps taken to identify single targeted clones (19). (B) Photograph of an undifferentiated D3 colony following eight days of selection on gelatinized plates in medium containing LIF. The pipette used for picking the colony is shown on the *right*. Magnification, 60 ×.

4.1 Screening for homologous targeting events using the polymerase chain reaction

The basic strategy is to design PCR primers that will amplify a novel junction fragment created by the correct homologous recombination event (see Chapter 1 for details). This can be done most easily for replacement-type vectors by having one of the homologous arms approximately 600–1200 base pairs in length (short arm). As illustrated in *Figure 3* the primers for the PCR reaction are then designed in such a way, that one primer binds to the *neo* cassette, and the second to a region just past the short arm of the targeting vector within the endogenous locus. To test for PCR conditions that allow for the identification of a homologous recombination event, we make control vectors which include the second primer annealing site (*Figure 3*) and establish ES cell lines containing one copy of this control vector. We then perform a series of control experiments with these cell lines to optimize the PCR reaction conditions for detecting single homologous targeting events in pools of negative cells. Approximately half of a colony of control ES cells is mixed with different amounts of wild-type ES cells to determine the pool size under which the rare homologous events can be detected.

Protocol 10. Screening procedure using the polymerase chain reaction (*Figure 4*)

Materials

- Pasteur pipettes
- dissecting microscope
- mouth pipette (Chapter 3, *Figure 3*) or pipetteman
- reaction tubes (500 μl)

Solutions

- PBS without Mg and Ca (Section 2.2)
- mineral oil
- proteinase K (10 mg/ml, stored frozen in aliquots)
- dNTP frozen (−70°C) stock solution containing 10 mM each of dATP, dCTP, dGTP, and dTTP

Protocol 10. *Continued*

- 10 × PCR buffer (100 mM Tris–HCl pH 8.3, 500 mM KCl, 15 mM $MgCl_2$, 0.1% gelatin)[a]
- *Taq* polymerase (Cetus)

Method

1. After about eight days of drug selection (*Protocol 8*), circle the colonies on the bottom side of the plate by holding the plate up to the light.
2. Replace the medium with PBS.
3. Pick half of each colony in a pool (or approximately equal cell volumes from each colony) with a drawn out Pasteur pipette (heat Pasteur pipette with a Bunsen burner, pull one end quickly, and break pipette at appropriate diameter, see *Figure 4*) and return cells to incubator.
4. Pool between 20 and 50 colonies, depending on the sensitivity of the PCR reaction, in one reaction tube.
5. Centrifuge the cells for 15 sec in a microcentrifuge, and then aspirate the PBS, leaving 5 μl.
6. Add 35 μl of ddH_2O, resuspend cell pellet by vortexing.
7. Put the tubes on dry-ice for 5 min.
8. Heat the tubes for 10 min. at 95°C.
9. Add 50 μl of mineral oil.
10. Cool the reaction tubes to 50°C.
11. Add 1 μl of proteinase K (10 mg/ml).
12. Incubate for 90 min. at 50°C.
13. Heat inactivate the proteinase K reaction at 94°C for 10 min.
14. Add to each lysate at 4°C a PCR reaction cocktail containing:
    - 5 μl of 10 × PCR buffer
    - 1 μl dNTPs
    - 1 μl of each primer (1 μg/μl)
    - 0.5 to 1 μl *Taq* DNA polymerase
    - H_2O to a final reaction volume of 50 μl
15. Run the PCR reaction for 40 cycles as follows:
    - denaturing conditions: 94°C/30 sec.–1 min.
    - annealing: 50–62°C/2 min.
    - elongation: 60–72°C/2–7 min.
16. Analyse 10 μl of the sample on an agarose gel (for details see 20).
17. Blot DNA on to a membrane.

18. Hybridize the blot with a probe spanning the synthesized DNA fragment.

19. Wash the membrane and expose for 5 min.–1 h (control DNA bands and products of homologous recombination events should be visible after 15 to 30 min. exposure).

20. Pick half of each colony from the pools which are positive on the Southern blot and repeat the PCR reaction (steps 5–19) on each single half colony. Transfer the other half of each colony to a 12 mm dish (one well of 24 well plate) containing a monolayer of feeder cells.

21. When the positive result is confirmed by PCR on single colonies, expand these lines after tryplating (*Protocol 11*) once or twice, depending on cell density, and analyse by Southern blot analysis.

22. Freeze and store at least one vial in liquid nitrogen (*Protocol 7*) as soon as possible. Cells can be used for blastocyst injections (Chapter 4) once a confluent 12 or 35 mm plate is obtained.

[a] We routinely test 10 × PCR buffers containing 10, 12.5, 15, 17.5, and 20 mM $MgCl_2$.

4.2 Screening for homologous targeting events by Southern blot analysis

The use of isogenic DNA vectors has greatly increased the gene targeting frequency (21, and our own unpublished results). Using this type of vector, targeting frequencies from 1/10 to 1/100 seem to be quite common using the PNS selection scheme. With such high targeting frequencies and the progress which has been made in ES cell culture techniques and DNA isolation methods it makes screening by Southern blot analysis a very attractive and straightforward procedure (see *Figure 5*).

Protocol 11. Growing cells for Southern blot analysis

Materials

- dissecting microscope
- 24 well plates containing feeder cells (*Protocol 2* or *3*)
- ES cell medium (Section 2.2)
- trypsin/EDTA (Section 2.2)
- PBS without Ca and Mg (Section 2.2)

Method

1. Pick the drug resistant colonies that form after about eight days of selection with a drawn-out Pasteur pipette (*Figure 4*) or a yellow Gilson tip under a dissecting microscope.

Protocol 11. *Continued*

2. Transfer each colony on to a 24 well plate containing feeders and ES cell medium, and break the colonies up into about five pieces.
3. Change the medium the next day.
4. After two days of culture, tryplate the cells using the following procedure.
   (a) Aspirate the medium and wash the cells with PBS.
   (b) Add 75 μl of trypsin/EDTA and incubate 3–5 min. at 37°C, or until the cells lift off the dish.
   (c) Add 500 μl ES cell medium and break colonies up by pipetting up and down using a pipetteman (blue tip).
   (d) Change the medium 12–24 h later.
5. If the colonies have not grown to confluency in two to three days, repeat step 4.
6. If the colonies are grown to confluency in two days, trypsinize the cells and transfer half of the cells in each well to a well on two different 24 well plates; one gelatinized plate, duplicate, containing ES cell medium supplemented with LIF, and the other containing feeder cells, master plate. The LIF-containing plate is for preparing DNA and the other is to keep and freeze.
7. The next day change the medium on both plates and tryplate if necessary after two days.
8. The master plate should be frozen (see *Protocol 13*) and the duplicate plate should be used for DNA preparation (see *Protocol 12*).

Protocol 12. DNA isolation from ES cell colonies (modification of 22)

Solutions

- PBS without Ca and Mg (Section 2.2)
- lysis buffer (100 mM Tris–HCl, pH 8.5, 5 mM EDTA, 0.2% SDS, 200 mM NaCl, 100 μg/ml proteinase K)
- 1 × TE (10 mM Tris–HCl, pH 8.0, 1 mM EDTA)
- isopropanol
- 70% ethanol

Method

1. Aspirate the medium from each well on the duplicate plate when it is confluent and wash with PBS.
2. Add 500 μl lysis buffer to each well and incubate 3 h or overnight at 37°C.

3. Add 500 μl isopropanol to each well and shake for approximately 15 min. on an orbital shaker until a DNA precipitate becomes visible.
4. Carefully take the supernatant off each well and add 500 μl of 70% ethanol to the DNA pellet.
5. Shake the plate again for 15 min.
6. Transfer the DNA precipitate with a yellow tip into an Eppendorf tube containing 50 μl of 1 × TE.
7. Incubate for 30 min. at 60°C to evaporate traces of ethanol.
8. Dissolve DNA overnight at 4°C (DNA will be ready for restriction enzyme digestions).

For the restriction enzyme digestions use 15 μl of the DNA solution. We have used the following restriction enzymes successfully: *Eco*RI, *Bam*HI, *Hin*dIII, *Bgl*II, *Cla*I, *Xba*I, *Sac*I, *Not*I, and *Xho*I.

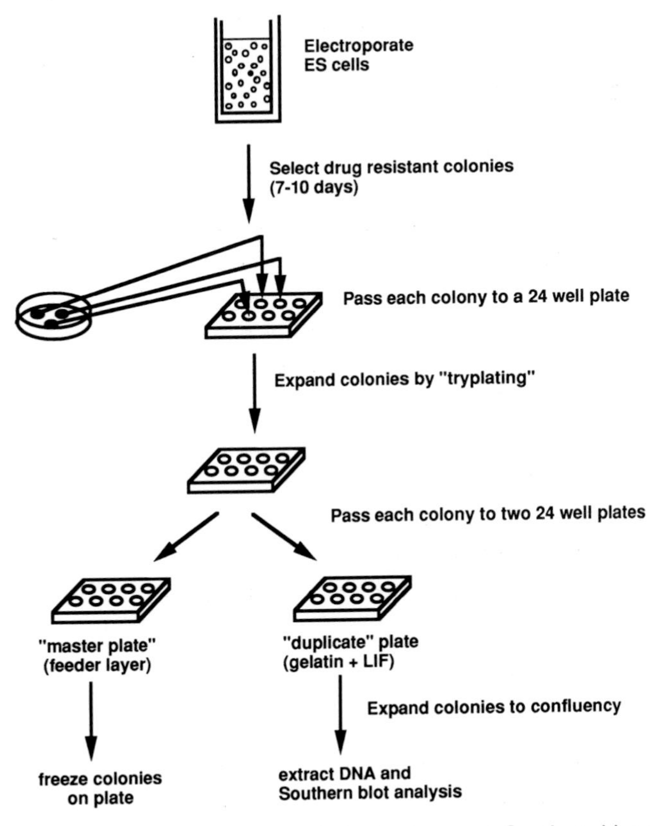

Figure 5. Strategy for screening for targeted colonies using Southern blot analysis.

4.3 Freezing of large numbers of genetically altered ES cell lines

Identification of homologous recombinants can require screening large numbers of *neo*^{R} ES cell colonies by Southern blot analysis. Such large numbers of ES cell lines can be frozen on feeders or on gelatin-treated 12 mm well plates (24 well plates).

Protocol 13. Short-term freezing of ES cell clones in 24 well plates (modified from 23)

1. Grow cells to confluency in 24 well plates containing feeder cells or gelatin (+LIF) in ES cell medium.
2. Put the plate on ice.
3. Remove the medium and add 0.25 ml 1 × cold freezing medium (see *Protocol 1*) per well on ice. Work quickly!
4. Transfer the plate to a pre-cooled styrofoam box (*Protocol 7*) in a −70°C freezer (the cells can be kept for at least six months).

Protocol 14. Thawing of ES lines frozen on 24 well plates

1. Remove the plate from the −70°C freezer.
2. Immediately add 1 ml of ES cell medium per well (work quickly!) and let it float in a 37°C water-bath.
3. When the ice crystals disappear, sterilize the outside of the plate by cleaning it with a tissue moistened with 70% ethanol.
4. Remove the medium from each well and add 0.5 ml ES cell medium to each well. If gelatinized plates have been used LIF has to be added to the medium.
5. If the ES cell colonies come off the plate in whole sheets, break them up by gently pipetting.
6. Change the medium the next day.
7. Passage the cells to feeder plates (35 mm) on the second or third day if cells have been grown to confluency. If not, tryplate them (*Protocol 11*) one or two times.

Acknowledgements

We wish to thank Malgosia Kownacka and Anna Auerbach for assistance with the feeder protocols, and Achim Gossler and Andras Nagy for suggestions on culturing D3 and R1 ES cells. This work was supported from grants from the Medical Research Council (MRC) and National Cancer Institute of Canada, and Bristol Myers Squibb to A.L.J. W.W. was partly supported by the Deutsche Forschungsge meinschaft. A.L.J. is an MRC Scientist and International Howard Hughes Scholar.

References

1. Martin, G. (1981). *Proc. Natl Acad. Sci. U.S.A.*, **78,** 7634.
2. Evans, M. J. and Kaufman, M. H. (1981). *Nature,* **292,** 154.
3. Robertson, E. J. (1987). In *Teratocarcinomas and embryonic stem cells: a practical approach* (ed. E. J. Robertson). IRL Press, Oxford.
4. Doetschman, T. C., Eistetter, H., Kutz, M., Schmidt, W., and Kemler, R. (1985). *J. Embryol. Exp. Morphol.*, **87,** 27.
5. Smith, A. G. and Hooper, M. L. (1987). *Dev. Biol.*, **121,** 1.
6. Smith, A. G., Heath, J. K., Donaldson, D. D., Wong, G. G., Moreau, J., Stahl, M., and Rogers, D. (1988). *Nature,* **336,** 688.
7. Williams, R. L., Hilton, D. J., Pease, S., Wilson, T. A., Stewart, C. L., Gearing, D. P., Wagner, E. F., Metcalf, D., Nicola, N. A., and Gough, M. M. (1988). *Nature,* **336,** 684.
8. Martin, G. R. and Evans, M. J. (1975). In *Teratomas and differentiation* (ed. M. I. Sherman and D. Solter), pp. 169–87. Academic Press, New York.
9. Ware, L. M. and Axelrad, A. A. (1972). *Virology,* **50,** 339.
10. Rudnicki, M. A. and McBurney, M. W. (1987). In *Teratocarcinomas and embryonic stem cells: a practical approach* (ed. E. J. Robertson) p. 19. IRL Press, Oxford.
11. McMahon, A. P. and Bradley, A. (1990). *Cell,* **62,** 1073.
12. Moreau, J.-F., Donaldson, D. D., Bennett, F., Witek-Giannotti, J., Clark, S. C., and Wong, G. G. (1988). *Nature,* **336,** 690.
13. Gough, N. M., Gearing, D. P., King, J. A., Willson, T. A., Hilton, D. J., Nicola, N. A., and Metcalf, D. (1988). *Proc. Natl Acad. Sci. U.S.A.*, **85,** 2623.
14. Nichols, J., Evans, E. P., and Smith, A. G. (1990). *Development,* **110,** 1341.
15. Lovell-Badge, R. H. (1987). In *Teratocarcinomas and embryonic stem cells: a practical approach* (ed. E. J. Robertson) p. 153. IRL Press, Oxford.
16. Zimmer, A. and Gruss, P. (1989). *Nature,* **338,** 150.
17. Potter, H., Weir, L., and Leder, P. (1984). *Proc. Natl Acad. Sci. U.S.A.*, **81,** 7161.
18. Joyner, A. L., Skarnes, W., and Rossant, J. (1989). *Nature,* **338,** 153.
19. Sedivy, J. M. and Joyner, A. L. (1992). *Gene targeting*. W. H. Freeman and Company, New York.
20. Sambrook, J., Fritsch, E. F., and Maniatis, T. (ed.) (1989). *Molecular cloning: a laboratory manual*. Cold Spring Harbor Laboratory Press, Cold Spring Harbor, NY.
21. te Riele, H., Maandag, E. R., and Berns, A. (1992). *Proc. Natl Acad. Sci. U.S.A.*, **89,** 5128.
22. Laird, P. W., Zijderreld, A., Linders, K., Rudnicki, M. A., Jaenisch, R., and Berns, A. (1991). *Nucleic Acids Res.*, **19,** 4293.
23. Chan, S. Y. and Evans, M. (1991). *Trends Genetic.*, **7,** 76.

3

Analysis of gene transfer in bone marrow stem cells

KATERI A. MOORE and JOHN W. BELMONT

1. Introduction

Pluripotent haemopoietic stem cells give rise to all the cells of the haemolymphoid system. Since haemopoiesis occurs in the bone marrow throughout adult life, genetic, neoplastic, and infectious diseases that affect this system could conceivably be treated by gene therapy targeted to haemopoietic stem cells (HSC). The possibility of gene therapy has focused efforts towards developing reliable methods for gene transfer into bone marrow. These efforts, inevitably, have raised many questions about stem cells and their role in the development and maintenance of haemopoiesis. One now sees a convergence of research goals in elucidating the biology of these fascinating cells and in gene therapy applications. The first demonstration that haemopoietic progenitors could be transduced with a retroviral vector was provided by Joyner *et al.* (1). In those experiments a vector carrying the bacterial Tn5, *neomycin phosphotransferase* (*neo*) gene was shown to confer G418 resistance to murine haemopoietic colonies. Williams *et al.* (2) and Dick *et al.* (3) extended these studies by demonstrating that very primitive precursors could also be transduced by similar vectors. In fact, the introduction of such neutral genetic markers provides some of the most rigorous evidence available for the existence of a transplantable stem cell.

The production of haemolymphoid cells (bone marrow cells; circulating blood cells; lymphoid cells of the thymus, spleen, lymph node, and epithelial surfaces; tissue macrophages; osteoclasts; and some CNS glial cells) is a dynamic developmental process. It begins in the early fetal period and continues throughout adult life. This process is organized as a developmental hierarchy proceeding from the most primitive, multilineage precursors to lineage-specific progenitors. In each cell this process follows an intrinsically determined developmental programme directed towards terminal differentiation. Committed progenitors are thought to have a continuously decreasing capacity for self-renewal although there are marked differences in the proliferative capacity and life span of the mature cells of each lineage. The processes of growth and differentiation are controlled by a complex extrinsic network of inductive factors including the haemopoietins, ligands for receptor tyrosine kinases, and cell adhesion molecules. Likewise, negative regulatory signals are provided by members of the TGF-β and MIP-α cytokine families. The context of the extrinsic signals (for example, cell developmental stage, micro-environment, interaction of sequential and simultaneous signals) determines the ultimate response of the cells.

At the root of the developmental hierarchy are a unique class of cells termed pluripotent haemopoietic stem cells. These cells can give rise to all the lineages of the haemolymphoid system, have huge capacity for proliferative differentiation, and are apparently capable of extensive self-renewal. They arise, in embryonic development, from the extra-embryonic mesoderm where the yolk sac blood islands form. They then migrate to the fetal liver, where haemopoiesis takes place during later *in utero* development, and finally take up residence in the bone marrow where they continue to survive in adulthood. Although these cells reside in discrete niches in the marrow, they can be removed and transplanted to autologous, syngeneic, or allogeneic recipients. From this observation, it can be inferred that they possess specific cell interaction molecules that allow them to home to an appropriate supportive micro-environment in the recipient marrow space (4). In both mice and humans post-transplant haemopoiesis can occur as the result of proliferation of a single stem cell (5, 6). In humans this would correspond to approximately 10^{15} daughter cells or 50 generations—demonstrating a unique capacity for proliferation and differentiation among adult somatic cells. Stem cells represent only one in 10^4–10^5 of the nucleated cells in unfractionated marrow (7). This fact alone presents a difficult problem for gene transfer, making efficient transduction of such a rare subpopulation dependent on near quantitative gene transfer. The potential difficulty of actual gene targeting in stem cells is further compounded by the often low frequency of homologous versus random integration events.

2. Methods for the transduction of haemopoietic stem cells

2.1 Physical methods of gene transfer

Two basic approaches for introducing DNA into mammalian cells have been developed; physical methods and viral vectors. The physical methods encompass a variety of techniques that include: electroporation (Chapter 2), lipofection, calcium phosphate co-precipitation, microprojectile, microinjection, protoplast fusion, and several others (8). A review of these techniques is outside the scope of this chapter. They may play a more important role in the future as applications not compatible with viral vector procedures are developed. Only electroporation has been used in published accounts of gene transfer into haemopoietic progenitors although the efficiency of gene transfer was 1–3% (9). In principle these methods allow the transfer of large segments of DNA containing all the necessary elements for precise regulation of gene expression. In addition, as outlined extensively in this book, targeted gene alteration likely depends on these physical methods of gene delivery.

2.2 Viral methods of gene transfer

The second class of gene transfer vehicles are based on viruses capable of infecting mammalian cells (10). These viruses use receptor-mediated pathways for the entry of the foreign genetic elements into target cells. In addition, they have specific mechanisms for the expression of their genomes either through maintenance of episomal elements or by integration into the host cell genome. Viral vectors are generally constructed with deletions in the virus genome that render them replication incompetent without the presence of helper or wild-type viruses. The complementing functions of the helper viruses can in some cases be fully substituted by replication-defective expression units that provide *trans*-acting viral proteins necessary for vector assembly and transduction. The viral vector systems that have been used most widely in mammalian cells include: SV40, adenovirus, herpes virus, adeno-associated virus, vaccinia, bovine papillomavirus, and retrovirus. Although adenovirus (11) and adeno-associated virus (12) vectors have been used to transduce marrow-derived cell lines, most studies to date have relied on retroviral vectors.

Retroviral vectors derived from the Type C Moloney murine leukaemia virus (Mo-MuLV) have been used extensively in animal and human models for gene transfer into haemopoietic precursors and lymphoid cells (13). One distinctive feature and advantage of retroviral vectors—efficient chromosomal integration (*Figure 1*)—appears to be essential in the haemopoietic system because of the extensive proliferation that occurs during the differentiation of stem cells (*Table 1*). Ecotropic retroviral vectors infect most rodent cell types due to the ubiquitous distribution of cellular receptors that mediate infection through envelope proteins. In addition, the host range can be altered by the use of amphotropic envelope proteins in packaging cell lines that can mediate infection of cells of non-rodent origin. This receptor-mediated cellular entry and an efficient viral mechanism for subsequent entry of the foreign nucleic acids into the target cell nucleus allows very high transduction efficiencies. In established tissue culture cells transduction approaching 100% is readily achieved. The percentage is lower in cultures of primary cells that are not rapidly dividing, largely due to the requirement of cell division for integration into the genome. The precise reason for this is unknown, although recent evidence suggests a requirement for breakdown of the nuclear envelope during mitosis to allow entry of the pre-integration proviral DNA complex (P. Brown, personal communication). Retrovirus

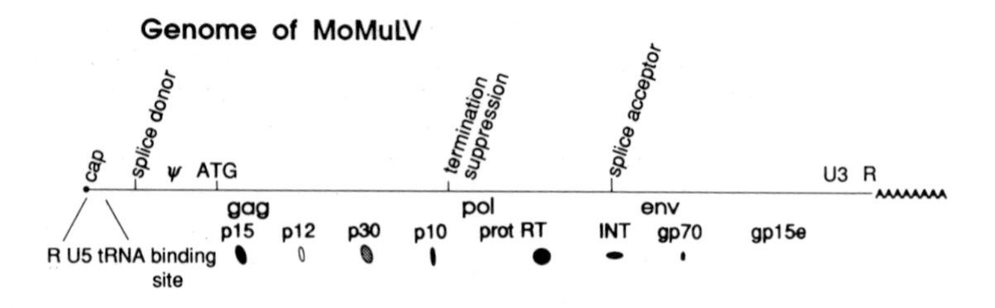

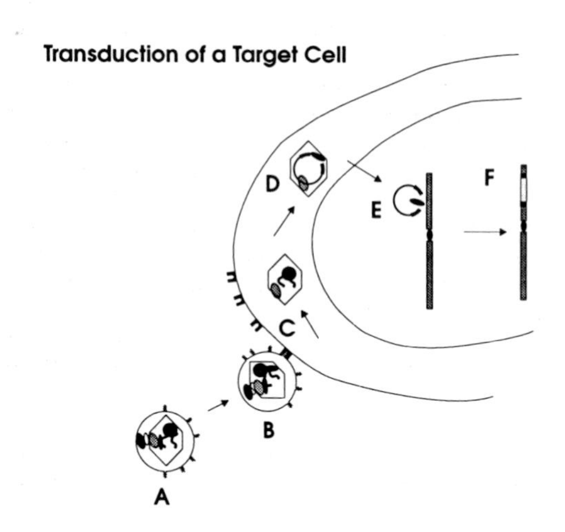

vectors have a number of disadvantages as well (*Table 1*). The most important disadvantage is a size limitation of approximately eight kilobases for insert DNA. This usually prohibits the inclusion of all regulatory elements necessary for tissue-specific or regulated expression.

The function of the packaging cell line used is crucial to successful gene transfer. To date, the packaging cell lines developed for these studies have all been generated in embryonic fibroblastoid cell lines, such as, NIH 3T3 cells. In the past the cell line ψ2 was most widely used for the production of ecotropic virus capable of infecting murine cells. This cell line contains a packaging plasmid element derived from Mo-MuLV that has a simple deletion of the ψ packaging region. This allows frequent recombination with vector sequences that produces replication competent virus. We prefer the third generation packaging cell lines developed by Markowitz *et al.* (14, 15). In these cells the packaging plasmids are divided into two separate expression constructs that separately produce the products of *gag-pol* and *env* genes. These packaging cell lines allow production of vectors free of helper virus in most cases.

Figure 1. Retrovirus lifecycle. The genome of each retrovirus particle consists of two identical genomic length messenger RNAs (*upper part of figure*). These RNAs are physically associated near the 5′ ends at a dimer linkage site. In the host cell the full-length RNA is translated to produce the *gag* (p15, p12, p30, and p10) and *pol* (prot—protease, RT—reverse transcriptase, and INT—integrase) gene products. Two polyproteins are translated—pr65gag and gp185gag-pol. Release of the component peptides occurs within the particle and is accomplished by the viral protease. The ratio of gag and gag-pol polyproteins is controlled by translation termination suppression at the boundary of the gag and pol coding sequence. The envelope gene products (gp70 and gp15e) are translated from a spliced message (splice donor, splice acceptor). The fixed ratio of gag-pol and env gene products is controlled by regulated splicing. In a retrovirus vector packaging cell line all these gene products are supplied in *trans* by two independent expression constructs. Viral or vector mRNAs that contain specific recognition sequences —ψ—are efficiently packaged into particles at the cytoplasmic membrane of producing cells. (A) The assembled particle contains two genomic RNAs associated with the nucleocapsid peptide (NC, p10) surrounded by the capsid protein (CA, p30). The p15 (matrix—MA) and p12 are necessary for assembly. The entire nucleocapsid is surrounded by host cell-derived lipid bilayer envelope studded with the envelope proteins. (B) The viral particle contacts the target cell via specific envelope–receptor interaction. The ecotropic envelope receptor is different from the amphotropic receptor. (C) After penetration of the viral particle reverse transcription is initiated using a cellular tRNA assembled with the particle in the producer cell as the primer (primer binding site—PBS). Regulated inter-strand jumps are required for reverse transcription of the genomic RNA to produce the double stranded provirus DNA. (D) The pre-integration complex which contains at least p30 and integrase peptides plus the provirus DNA is translocated to the nucleus. (E) Integration is mediated by the integrase protein which has endonuclease activity. (F) Proviruses are found at random positions in the genome of the target cell. Preference for open chromatin domains has been shown.

Table 1. Retroviral vectors

Advantages	**Disadvantages**
• Broad host range	• Random site of integration
• High efficiency transduction in cultured cells	• Inadequate control of expression
• Stable integration	• Produced by mammalian cells
• Helper virus-free	• Size limitation, insert $\leq$ 8 kb
	• Integration requires target cell replication

2.2.1 Retroviral vector design

Several reviews are available that describe the basic features of retrovirus vectors and the use of these vectors for expression of genes in haemopoietic cells (10, 13, 16). Several possibilities are shown in *Figure 2*. These include the following.

(a) Basic. A cDNA is expressed from the genomic or spliced transcript initiated in the Mo-MuLV LTR.

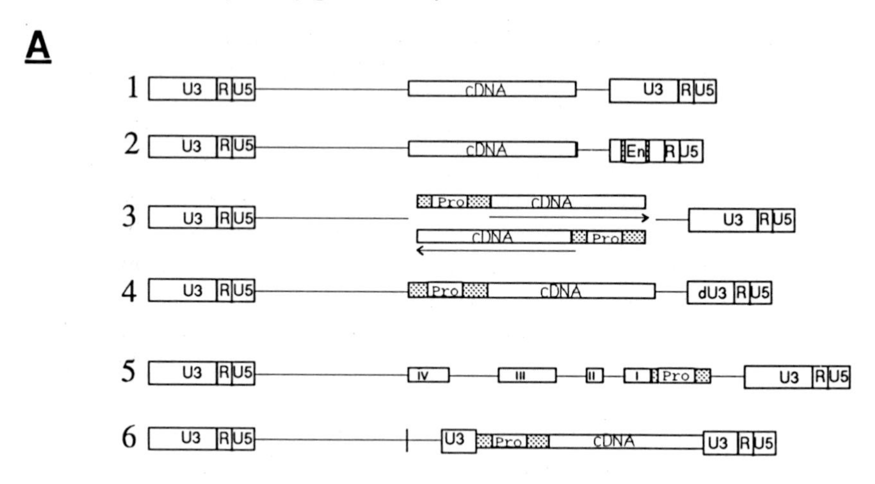

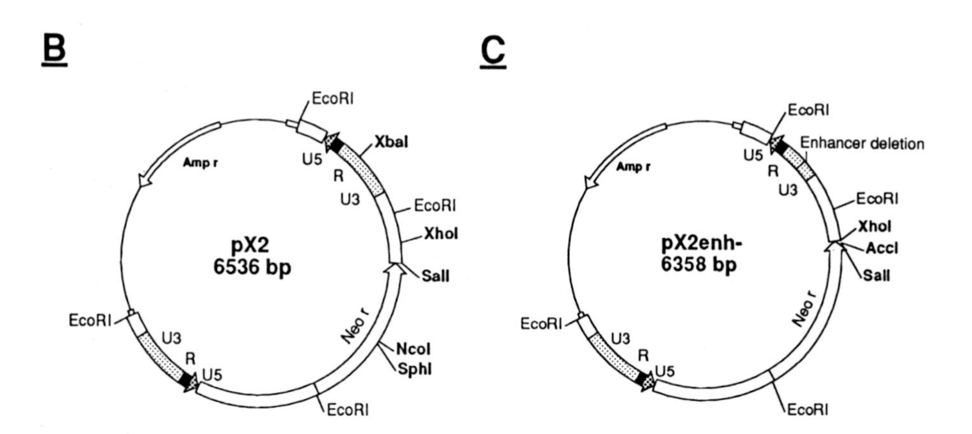

Figure 2. Retroviral vector design. (A) Major design categories used for expression of genes in marrow cells. 1—basic, LTR promoter; 2—LTR enhancer substitution; 3—independent promoter for the cDNA that can be cloned in either orientation; 4—deletion of the 3′ LTR enhancer or promoter, self-inactivating vector; 5—'minigene' containing introns for correct splicing and regulation, can be used with self-inactivating vector; 6—double copy, expression construct cloned into the 3′ LTR leads to duplication after integration. (B) pX2 and (C) pX2enh-—plasmid vector intermediates for expression constructs. pX2 contains unique cloning sites within the vector and in the 3′ LTR. (Designed and constructed by E. Aguilar-Cordova and F. Amaya-Manzanares. Available on request.)

(b) Enhancer substitution. Key transcription control elements of the Mo-MuLV LTR are replaced with those from other viruses or cellular genes. This has been useful in expressing genes in embryo-derived teratocarcinoma (EC) cells and embryonic stem (ES) cells (17).

(c) Internal promoter. These use an independent transcription unit (promoter and enhancers) for the cDNA. This can be placed in either the forward or reverse orientation with respect to the LTR and is compatible with the self-inactivating vector design.

(d) Self-inactivating. Deletions in the U3 region of the 3′ LTR are duplicated into the 5′ LTR of the target cell's integrated provirus because of the ordered nature of template RNA strand jumping during reverse transcription. The resulting provirus has transcriptionaly inactive LTRs. This allows for internal and possibly tissue-specific regulatory control for transcription of transferred sequences.

(e) Internal gene. These include promoter, enhancer, introns, and exons for the gene of interest. Size constraints on vector packaging limit the range of applications.

(f) Double copy. The cDNA transcription unit is placed in the 3′ LTR so that during vector replication and integration it is duplicated into the 5′ LTR (18).

We currently prefer smaller basic vectors for applications where LTR-controlled expression is sufficient. In practice, the performance of individual designs is rather unpredictable. Most laboratories have constructed sets of vectors incorporating a variety of design strategies and tested them empirically. In addition to other concerns, design of vectors for expression in stem cells is limited by the necessity for very high titre production. Efficient transduction requires titres of at least 10^6 units/ml, because of the low frequency of stem cells in unseparated populations, low receptor density, low rate of replication, or other unknown factors. Secondly, data from several laboratories indicates that lower and more variable levels of expression are obtained when using vectors that contain two transcription units as is the case with vectors carrying a selectable marker (13). Improved vector production will be an important component of successful commercialization. These improvements may include: new packaging cell lines, perhaps, based on cells adapted for bioreactor growth, improved vector collection, concentration, storage, and quality control.

2.2.2 General considerations

(a) Vector plasmid is prepared by double CsCl banding (other techniques, e.g. gel purification are also acceptable).

(b) Packaging cell lines, GP + E-86 (GPE, ecotropic) and GP + *env*-AM12 (AM12, amphotropic) can be obtained from Dr A. Bank (Columbia University, New York).

   i. Upon receipt of cells the cultures should be expanded. At least 20 vials containing 10^7 cells in regular media plus 7.5% DMSO (cryomedia) are frozen and maintained in liquid nitrogen. These early passage stocks are important for the consistent production of vector since the packaging cell lines are somewhat unstable in the production of virus when maintained in continual passage.

ii. The packaging cell lines may also be reselected for the packaging plasmid *gpt* marker in HXM media; 15 μg/ml hypoxanthine, 250 μg/ml xanthine, and 25 μg/ml mycophenolic acid. The AM12 cells are further selected in hygromycin B 200 μg/ml, for the presence of the hyg^R gene.

iii. Upon isolation of any clone of vector producer cells this expansion procedure should be repeated. New vials from early passage stocks should be used for each experiment to maximize consistency. The vector producer cells should be discarded after four to six weeks in culture.

(c) Titring of virus-containing supernatant from cell lines can be accomplished by several methods.

i. For the conventional N2-type vector (*Figure 2A*, number 1 and 2) the titre after transfection from the population will be from 10^5–10^6 $G418^R$ c.f.u./ml.

ii. For non-selectable markers one assays for vector RNA or proviral DNA. Vector RNA can be measured by dot-blot hybridization of supernatant extracts, RNA PCR on the supernatant, Southern analysis on transduced target cells, or PCR on the transduced target cells. All of these can be quantitated by serial dilution of the vector-containing supernatant. PCR on transduced target cells is our preferred method as it is easily performed and standardized compared to the other methods.

iii. Direct detection of the gene product can be accomplished by specific assays for the protein in transduced target cells. If monoclonal or polyclonal antibodies to the protein exist, immunocytochemistry offers a method for determining active vector titre at the single cell level (19). Other assays include zymogram, flow cytometry, ELISA, immunoprecipitation, and western analysis. These generally lack the sensitivity necessary for titring but may be useful for screening clones.

iv. Multiple examples of the correlation of expression of the transfected construct and the viral titre have been made (13). This apparently indicates a common pool of translatable and packageable mRNA and indicates the importance of transcription in the packaging cell to vector titre. Screening methods based on expression of the vector in the packaging cell line are usually successful in identifying clones with high titre.

(d) Amplification of titre may be necessary if initial transfected producer population titres are too low (less than 10^6). This can be accomplished by identifying individual clones of the vector-producing cell line, or by serial reinfection of packaging cells by virus from producer populations of the alternate envelope specificity or tropism, (i.e. ecotropic or amphotropic).

i. Ecotropic and amphotropic packaging cells are resistant to reinfection by vector particles of the same tropism due to cell surface receptor interference; but they can be reinfected by vector of the other tropism.
ii. For vectors with intact 3′ LTRs serial reinfection is a very effective method for increasing vector production. This results in a large increase in copy number of the vector in the packaging cell line and a comparable increase in packageable mRNA.
iii. In general, we have been more successful increasing titres in amphotropic producers, but we have not been able to increase the titre beyond 5×10^7 for either tropism.
iv. For vectors which contain alterations in the 3′ LTR, (e.g. enhancer deletions or double expression constructs) only the cloning method can be used since only the primary transfectants produce virus in a packageable form.

Protocol 1. Transfection of the retrovirus vector

1. Trypsinize subconfluent packaging cells in active growth phase, wash once in media, enumerate in a haemocytometer, and resuspend in media at 1×10^7 per ml.
   (a) trypsin-EDTA; 0.05% trypsin, 0.053 mM EDTA-4Na in 0.85% NaCl (Gibco, Grand Island, NY)
   (b) standard growth media:
   - Dulbecco's Modified Eagle's Media (DMEM, Gibco)
   - 10% bovine calf serum, defined iron-supplemented (BCS, Hyclone Laboratories Inc., Logan, Utah)
   - antibiotics, e.g. gentamicin or penicillin–streptomycin, optional (not routinely used for established cultures)
2. Place 0.35 ml (1×10^7 cells/ml in media) in the electroporation cuvette, mix with 40 μg of the vector plasmid and when necessary 4 μg of a selectable marker construct, (e.g. pSV2*neo*, ATCC), and pulse using a BioRad Gene Pulser (BioRad, Richmond, CA) at 280 V and 500 μF. Immediately place the cells on ice and process the next sample.
3. After all the vectors are electroporated, mix the cells from each with 10 ml of media and seed them into a new T75 flask (75 cm^2, Corning, Corning, NY). After 24 h remove non-adherent, dead cells and add fresh media.

Protocol 2. Selection of virus-producing cells transcribing *neo*

1. Place the cells in selective media 48 h after electroporation.
   (a) Selection media; standard growth media plus 400 μg/ml G418 (active concentration, Geneticin^{tm}, Gibco).

Protocol 2. *Continued*

(b) If working with double copy vectors or vectors with alterations in the 3′ LTR, distribute cells into five to ten 100 mm dishes 48 h after electroporation, for direct cloning (see *Protocol 5*).

2. Feed these cultures every three days with G418 media for approximately 10–14 days. At this time the non-transfected cells will die.
3. Trypsinize the selected cells and establish a pooled population for expansion and cryopreservation.

Protocol 3. Determining titre for vectors with the selective marker *neo*

1. Grow vector producing cells to 80% confluence in a T25 or T75 flask. Remove selective media, wash twice, and replace with 5–10 ml fresh growth media.
2. Seed 6 well trays with 1×10^4 NIH 3T3, Rat208F, or other suitable target cell line per well in growth media.
3. After 24 h remove the culture supernatant from the producer cells, filter through 0.4 μm cellulose–acetate syringe filter (Nalge Company, Rochester, NY).
   (a) This supernatant may be frozen at −70°C without significant deterioration in bioactivity.
   (b) Titre is reduced by multiple rounds of freezing and thawing.
4. Starting with 1 ml, make serial ten-fold dilutions with media and virus supernatant in the tray of target cells. Add polybrene (Sigma, St. Louis, MO) to final concentration of 4 μg/ml from 100 × stock in water (stored frozen). Make dilutions to 10^7 and include a no virus control.
5. After 24 h replace vector-containing media with fresh growth media. After an additional 24 h, replace with selective media.
6. Aspirate media and refeed cultures every three days.
7. After two weeks stain with methylene blue (0.1% methylene blue, 50% methanol) and enumerate individual colonies.
8. Titre is expressed as the number of colonies present in the well with the highest dilution, (e.g. if three colonies are present in the 10^6 well, then the titre is 3×10^6 G418^R colony forming units/ml (c.f.u./ml)).

Protocol 4. Titration of a vector without a selectable marker or standardization of titration by PCR

1. Perform steps 1–5 in *Protocol 3*, except maintain the cells in growth media.

2. After 72 h trypsinize the cells from the target cultures. Wash twice with D-PBS (Dulbecco's phosphate-buffered saline without Ca^{2+}, Mg^{2+}, Gibco).
3. 'Quick-prep' method for preparation of DNA.
   (a) To cellular pellet in 1.5 ml microcentrifuge tube, add 100 µl quick lysis buffer (*Table 2*), mix well.
   (b) Incubate overnight at 37°C, or 4 h at 55°C.
   (c) Heat inactivate, 85–90°C 10–15 min.
   (d) Centrifuge 5 min., microcentrifuge.
4. Use one tenth of this material as substrate for vector-specific PCR. Primer choice should be individualized and optimal conditions for PCR established before application to this procedure. Maximal sensitivity detection is required so attention should be given to appropriate negative controls. In each experiment these should include sets of target cells cultured and processed in parallel but not exposed to vector containing supernatant.
5. PCR products may be analysed by simple agarose gel electrophoresis. To maximize sensitivity and specificity, Southern blot hybridization with a probe derived from sequences internal to the PCR oligonucleotide may also be used.
6. Titre is determined as the highest dilution that confers a detectable PCR product. This usually correlates well with a $G418^R$ titre.

Protocol 5. Cloning of high titre vector-producer cell lines

1. Seed five to ten 100 mm dishes with 10^3 cells after selection or (if working with double copy vectors or vectors with alterations in the 3′ LTR) from electroporated cells; maintain in culture.
2. Identify individual clones under inverted microscopy and isolate 10–20 by ring or swab cloning.
   (a) Ring cloning.
      i. sterile 5 mm plastic rings are dipped in autoclaved vacuum grease and placed on the surface of the dish around the clone.
      ii. Fill the ring with trypsin (< 200 µl), after 5 min. aspirate the fluid, and pipette into 1 ml media/well of a 24 well tray (one clone/well). Repeat. Examine and replace media after 24 h.
   (b) Swab cloning.
      i. Gently scrape a discrete isolated colony with a sterile cotton swab pre-moistened with media.
      ii. Rinse the swab in 1 ml media/well in a 24 well tray (one clone/well). Repeat. Examine and replace media after 24 h.

Protocol 5. *Continued*

3. When the wells are confluent, trypsinize and transfer the cells to a 6 well tray and eventually expand to T25 flasks.
4. When the cells are approaching confluence in T25 flasks, harvest supernatant and titre as above.

Protocol 6. Amplification of vector titre by serial reinfection

1. Collect 100 ml supernatant as above from transfected populations of ecotropic and amphotropic packaging cells. Combine, filter (0.45 μm), and freeze in 10 ml aliquots to generate adequate stock for entire procedure.
2. Seed a T75 flask with fresh packaging cells of the desired tropism at 20% confluence.
3. Twenty four hours later begin infection of the packaging cells by adding 10 ml undiluted supernatant from the transfected producers of the other tropism with polybrene at 4 μg/ml.
4. After 4–8 h remove media, replace with 10 ml growth media for overnight.
5. Repeat this procedure daily for up to ten days.
   (a) Split cells as they reach confluence. They will be ineffectively transduced if not growing well.
   (b) Repeated media changes and exposure to polybrene are disruptive. A rest interval of two to three days after three or four rounds of infection may be necessary to re-establish the health of the culture.
6. After the serial infection is complete, cell populations are titred. Individual high titre cell lines may be isolated by cloning.

2.2.3 Replication-competent virus

It is critical to monitor cultures for recombinant replication-competent virus (RCV) production. These viruses will cause lymphoma in bone marrow transplant recipients and confuse the interpretation of expression studies. Amphotropic Moloney viruses have not been found to cause disease in normal humans. However, the recent observation of lymphoma in rhesus monkeys transplanted with marrow exposed to amphotropic RCV suggests caution on the part of laboratory workers handling these viruses. Biological Safety Level 2 (BSL2, NIH Guidelines) containment is recommended for this purpose. Several methods can be used to detect replication-competent viruses of different tropisms. These include the XC plaque assay, S+L− assay, marker gene rescue assay, and PCR. We prefer the latter two because of their ease and sensitivity. Biosafety considerations for clinical applications are

Table 2. Buffers for PCR analysis

1 × PCR buffer	**Stock**	**10 ml 10 × buffer**
100 mM Tris–HCl pH 8.3	1 M Tris–HCl pH 8.3	1.0 ml
500 mM KCl	2 M KCl	2.5 ml
25 mM $MgCl_2$	1 M $MgCl_2$	0.25 ml
0.1% gelatin	2% gelatin	0.50 ml
	water	5.75 ml

Red cell lysis buffer	**Stock**	**50 ml**
0.32 M sucrose	1 M	16.0 ml
10 mM Tris–HCl pH 7.5	1 M	0.5 ml
5 mM $MgCl_2$	1 M	0.25 ml
1% Triton X-100	100%	0.50 ml (w/v)
	water	32.75 ml

Quick lysis buffer	**Stock**	**10 ml**
10 μl	10 × PCR buffer	1.0 ml
10 μl	1 mg/ml proteinase K	1.0 ml
10 μl	34 μM SDS	1.0 ml
10 μl	400 mM DTT	1.0 ml
60 μl	water	6.0 ml

All solutions are prepared in ddH_2O and autoclaved.

beyond the scope of this chapter, but these applications require much more extensive testing for adventitious viruses. In the marker rescue assay, a cell line (SVBNEO) produced by transfection of NIH 3T3 cells with a replication-defective *neo*-transducing retrovirus vector, is used as the rescue detector and is available from the authors. Infection of SVBNEO cells with either ecotropic or amphotropic RCV results in restored competence of the *neo*-transducing retrovirus. *Neo* vector production is then detected by G418 selection after infection of virgin NIH 3T3 cells. The marker rescue assay is sensitive, but requires at least one month to complete. To shorten the time period and allow a simpler screening method for new producer cells, an assay based on PCR has been devised (20). In this assay the oligonucleotide primers amplify Moloney sequences between the 3′ end of *pol* and the 5′ end of *env*. The specificity of the *env* sequences allows distinction between ecotropic and amphotropic tropisms. In the presence of Moloney-related RCV, these primers will amplify a fragment of 500 base pairs with ecotropic primers and 700 base pairs with amphotropic primers. Supernatants from the packaging cell lines themselves always contain inappropriately packaged cellular RNAs and DNA from the packaging elements which are detected by the PCR. Therefore, DNA from test transduced indicator cells is used, while the producer cell lines provide appropriate positive controls.

Protocol 7. Marker gene rescue assay

1. Seed 2×10^5 SVBNEO cells into a T25 flask.
2. Collect supernatant from vector-producing cell lines as for titring.
   (a) For negative control include supernatant harvested from SVBNEO cells.
   (b) For positive control collect supernatants from wild-type Mo-MuLV or known positive retrovirus producer cells.
   (c) Serum or plasma (10 μl) from experimental animals can be used to detect *in vivo* RCV.
   (d) Media supernatants from transduced bone marrow cultures can also be used in this assay.
3. Use 5 ml of supernatant to 'infect' the SVBNEO cells in the presence of 4 μg/ml polybrene. Serum or plasma is added directly to 5 ml growth media.
4. After 24 h change to growth media and maintain the cultures for two weeks. Expand to T75 flasks when confluent. Maintain one flask per test, splitting as necessary to keep cells in active growth.
5. After two weeks, collect 10 ml supernatant from the SVBNEO cells, filter (0.45 μm), and add to 5×10^5 NIH 3T3 cells in a T75 flask. Add polybrene to a final concentration of 4 μg/ml.
6. Remove media after 24 h and replace with fresh growth media; after an additional 24 h replace with and maintain cultures in G418 media for two weeks.
7. The appearance of G418 resistant clones indicates the presence of RCV in the initial inoculum.

Protocol 8. PCR for recombinant ecotropic and amphotropic virus: RCV-PCR

1. Seed 2×10^5 NIH 3T3 cells into a T25 flask.
2. Infect with 5 ml 'test' culture supernatants as in step 3 of *Protocol 7*. Plasma and media from test animals and cultures may also be used as above. DNA from animal tissues may be used directly in the PCR assay.
3. After 24 h change to regular growth media. Expand and maintain the cultures until they reach confluence in a T75 flask.
4. Harvest the cells by trypsinization, wash twice with D-PBS, and prepare DNA by standard phenol/chloroform extraction methods.

5. Oligonucleotide primers for RCV-PCR.
   - amphotropic primers:
     5′GAACCATCAAGGAGACTTTAACTAAATTAA3′
     5′AAGGCATCTTGTACAGTTCCCAGGAGGGAG3′
   - ecotropic primers:
     5′CAAGAGTTACTAACAGCCCCTCTCTCCAAG3′
     5′TGGTTGCCAGAAGTTGCCCATACCGTCTCC3′
6. Assay conditions for RCV-PCR.
   - 1 × PCR buffer (*Table 2*)
   - 0.2 μM each primer, either amphotropic or ecotropic pairs
   - 0.25 mM each dNTP
   - 1 U *Taq* DNA polymerase
   - 100–200 ng DNA
   - water to volume, (usually 50 μl reaction volume)

   (a) Denature 5 min. at 95 °C. Cycle for 30 rounds; 95 °C 1 min., 72 °C 1 min.

   (b) Electrophorese 10–25 μl reaction mixture through 1.3% agarose gel; visualize bands by ethidium bromide staining.

---

3. Tissue culture models of murine and human haemopoietic precursor cell maintenance, growth, and differentiation

3.1 Inductive signals for proliferation, differentiation, and self-renewal

Over the past several decades tissue culture models have been developed that demonstrate the long-term maintenance and clonal growth of haemopoietic precursor cells *in vitro*. These culture systems have allowed the delineation of the developmental hierarchy of haemopoietic cells. They have also served as assays for the growth of colony-forming cells (CFC) and have thus allowed the identification, purification, and cloning of numerous cytokine regulatory proteins, specific for growth of these colonies. Evidence from these experimental models points to the importance of the context of growth factor or cytokine stimulation, with respect to interaction with other factors present and the history of the individual cell. Progenitors appear to express several of the known cytokine receptors and presumably there are many other ligands and receptors that regulate their growth. Lineage determination is partially controlled by the stimuli to which cells are exposed. Some of the cytokines also function as trophic factors in whose absence the target cell undergoes apoptosis. Comparable information on the pluripotent stem cell might lead to culture conditions for its propagation without differentiation—a key goal in gene therapy research. This is likely to be essential for gene targeting strategies of gene therapy.

3.2 Colony assays for mouse and human progenitors

In the mid-1960s, the development of culture conditions that allowed the clonal growth of murine bone marrow cells in semi-solid medium was reported (21, 22). Later on, the cloning of human haemopoietic cells was described by Fauser and Messner (23). Growth of these cells required a soluble additive present in the conditioned media from cultured tumour cells or primary haemopoietic cells stimulated with mitogens or lectins. The term colony-stimulating factor (CSF) was coined to denote this additive and soon many other glycoprotein molecules with activities on haemopoietic cells were identified (24). The 'haemopoietic growth factors' are found in conditioned medias from a variety of sources including marrow stromal cells, tumour cell lines from a variety of tissues, and haemopoietic cells, especially T cells. These molecules have been shown to have specific actions on certain cell types and some show synergistic activity when used in various combinations. A discussion of this complex set of interactions is beyond the scope of this chapter (see reference 24 for review).

Colony assays provide evidence for the presence of progenitor cells that are able to produce single and multilineage colonies of differentiated cells. These assays have proven useful in analysis of gene transfer efficiency in both mouse and human experiments. Growth factors associated with progenitor cell growth are well known in both of these systems. They also allow assessment of stem cell maintenance in long-term bone marrow cultures as shown by cellular outgrowth of clonogenic progenitors. Individual haemopoietic colonies can be analysed by PCR for proviral integration and for expression by other means to reflect retrovirus infection efficiency. For example, cells transduced with a *neo* containing retrovirus can be selected in G418 for expression of the provirus. Other assays applicable to study of provirus expression in individual colonies include: analysis of isolated mRNA by reverse-transcription to cDNA and subsequent PCR (RT-PCR), micro-enzyme analysis, and immunocytochemistry on cytospin slides.

Further modifications of the basic colony assay have been developed that allow assessment of more primitive or non-committed progenitor cell populations. Blast cell colony assays (CFU-Bl) for both mouse (25) and human (26) have been described. The cells of these colonies are morphologically immature and, when replated give rise to secondary colonies of various lineages including more blast cell colonies. The high proliferative potential colony-forming cell assay (HPP-CFU) detects very large colonies of greater than 0.5 mm diameter at 10–12 days of culture in murine marrow after 5-fluorouracil (5-FU) treatment (27). We have concentrated on the use of the basic colony assay for evaluation of gene transfer efficiency, especially in the analysis of the long-term culture-initiating cell (LTC-IC) explained below.

We encourage perusal of the literature in this area for a more complete treatment of the methodology (28, 29).

3.2.1 General considerations

(a) Methylcellulose (MC) as the semi-solid support medium for colony assays has several advantages. Foremost is the relative ease of picking colonies for replating, DNA preparation, or microscopic examination. Methylcellulose is less viscous at room temperature than at 37°C but requires volume displacement pipetting for accuracy. Methylcellulose preparations for both mouse and human progenitor assays are available from Terry Fox Laboratory, Vancouver BC.

(b) Fetal calf serum and BSA selection and testing is very important. Test different lots and select one which achieves the best growth with a standard media base.

(c) Preservative-free heparin should be used in collecting human marrow samples.

(d) Colony numbers vary greatly depending on the proliferative status of the marrow, experimental manipulation, and growth factor combinations. An average frequency of clonogenic progenitors is one in 1000 nucleated bone marrow cells.

(e) Use only double-distilled water (ddH_2O).

(f) Plasticware is preferred for pipettes and other supplies. All glassware should be thoroughly rinsed with ddH_2O prior to sterilization. Low level contamination with organic solvents or soaps will cause poor cell growth.

Protocol 9. Murine cell preparation

1. Sacrifice mice by cervical dislocation, saturate with 70% ethanol.
2. Secure mouse to dissection board, reflect skin from the hind limbs, and flood with 70% ethanol.
3. Reflect muscles at their attachments, cut away as much as possible with scissors.
4. Separate femurs and tibias from the body at the hip joint, remove feet.
5. Remove as much muscle as possible then place the leg in a dish of IMDM (Iscove's modified Dulbecco's media, Gibco) with antibiotics on ice.
6. Once all bones have been collected, transfer to fresh dish of media and transport to tissue culture hood.
7. Remove bone epiphyses with scissors and using a syringe (25–27 gauge needle) filled with IMDM plus 10% BCS, expel the marrow.
8. Prepare single cell suspension by gentle repeated pipetting. Perform viable cell count by trypan blue exclusion. $2–4 \times 10^7$ bone marrow cells can be obtained from one mouse.

Protocol 10. Human marrow cell preparation

1. Obtain the light density fraction (<1.077 g/ml) from bone marrow buffy coat over standard Ficoll gradient (Ficoll-Pague^{tm}, Pharmacia, Piscataway, NJ).
2. Alternatively enrich for clonogenic progenitors by Percoll density gradient separation. Establish a 40/60 Percoll gradient from a stock of 100% Percoll (Sigma) (30). It is important to adjust the 100% Percoll solution to a refractive index of 1.3521 ± 0.0024 to obtain the correct density in the diluted samples.
   (a) Make dilutions with D-PBS.
   (b) Prepare gradients by layering 15 ml of 40% solution over 15 ml of the 60% solution in a 50 ml conical tube (Corning).
   (c) Layer with the buffy coat cell suspension; 1×10^9 nucleated cells (from 30 ml marrow sample) in 10–15 ml 2% FCS D-PBS.
   (d) Centrifuge at 850 *g* for 20 min.
   (e) Collect the cells at the 40/60 interface and wash twice in 2% FCS D-PBS.
   (f) Perform viable cell count; yield, approximately 1×10^8 mononuclear bone marrow cells (BMC) if starting with 1×10^9 nucleated cells. Expect a six-fold enrichment in clonogenic progenitors upon colony assay.
3. Alternatively enrich for clonogenic progenitors by immunopanning. Immunoadherence panning for $CD34^+$, soybean agglutinin positive (SBA^+) cells can be accomplished with AIS MicroCellector^{tm} T25 cell culture flasks (Applied Immune Sciences, Menlo Park, CA). See the packaged insert for the detailed protocol. Expected enrichment, compared to standard Ficoll preparations, is 10–50-fold.

Protocol 11. Preparation of 2% stock methylcellulose

1. Prepare 2 × IMDM by dissolving a one litre packet plus 3.25 g $NaHCO_3$ in a total volume of 500 ml water. Sterilize by filtration through a 0.22 μm filter.
2. Add 20 g methylcellulose (Methocel MC 4000 mPa.s, Fluka Chemical Corp., Ronkonkoma, NY) to 500 ml boiling ddH_2O in weighed, sterile two-litre Erlenmeyer flask. Reheat just to the boiling point and remove immediately from heat.
3. Cool to 40–50°C under running water and add the 500 ml of double strength IMDM. Hold overnight at 4°C with continuous stirring. Check weight, adjust to 1000 g with sterile water if necessary.

4. Dispense in 40 ml aliquots in 50 ml conical tubes. These can be stored at −20°C indefinitely. Avoid frost-free freezer. Thaw at room temperature.

Protocol 12. Culture preparation

1. Prepare methylcellulose mix by combining the following (final concentration):
   - prepared MC stock (0.8–0.9%)
   - fetal calf serum (30%)
   - deionized bovine serum albumin (1%)
   - 2-mercaptoethanol, tissue culture grade (10^{-4} M)
   - L-glutamine (0.29 mg/ml)
2. Prepare colony-stimulating factors as follows. 3 U/ml erythropoietin and 20 U/ml murine IL-3 (Genetics Institute) for murine cells, or 5% PHA-LCM (phytohaemagglutinin-stimulated lymphocyte conditioned medium, Terry Fox Laboratory) for human cells. The conditioned media contain several growth factors including IL-3 and GM-CSF.
   (a) For murine assays, we have also used conditioned medias (CM) from WEHI-3B and 5637 (ATCC) cells as a source of CSF at 10% each, and lowered the FCS to 20%. 5637 may also be used in place of PHA-LCM in human assays.
   (b) Recombinant growth factors available commercially can be used as recommended (Collaborative Biomedical Products or see the Linscott Catalogue for alternative sources).
3. Dilute cells in IMDM plus CSF mixture to give 5×10^4-1×10^5 mouse BMC, or 2×10^5 human BMC per 1 ml culture.
4. Prepare cultures in triplicate. Mix cells, supplements, and methylcellulose in a 3 ml syringe fitted with an 18 G needle. Avoid bubbles.
5. Dispense 1 ml in the middle of a 35 mm suspension culture dish, rock gently or rotate to spread. Dishes from different suppliers vary as to adherence promoting activity. Greiner and Sarstedt are recommended.
6. Place two dishes of cells plus one dish of sterile water in 10 cm Petri or tissue culture dish (*Figure 3*). Transfer to incubator, fully humidified, 5% CO_2, 37°C.
7. Examine at 10, 14, and 21 days with an inverted microscope.

3.2.2 Colony identification

Colonies are defined as greater than 50 cells, although their size varies widely. Colonies of more than 50 000 cells have been reported but an average-sized

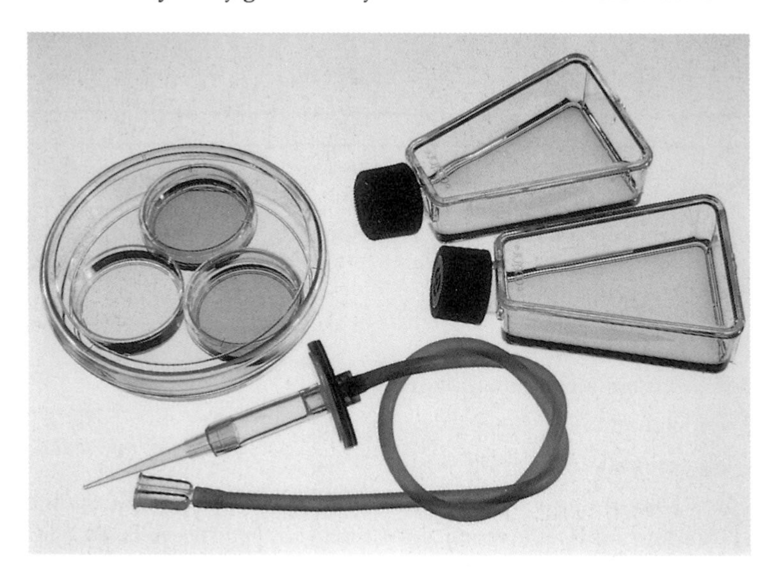

Figure 3. Experimental apparatus used in haemopoietic cultures. Long-term bone marrow culture (T25 flasks, Costar, 0.2 μm vented filter cap); colony assay (two 35 mm test dishes are placed in a 100 mm dish with a dish of sterile water); and picking of individual CFC (a 0.2 μm syringe filter is fitted to a mouthpiece and to tubing that will accommodate a micropipette tip on the opposite end).

colony has 200–500 cells. Size is dependent on the proliferative status of the marrow and the combination of growth factor additives. With practice, individual colony-types can usually be determined by microscopic inspection (*Figure 4*). Confirmation can be accomplished by individual colony picking and microscopic inspection of cytospin slide preparations stained with Wright-Giemsa stain or Dif Quik^{tm} (*Figure 4*). The following is a list of common colony-types and suggested stains for positive identification.

(a) BFU-E. Erythroid 'burst'-forming unit, erythropoietic progenitors more primitive than progenitors for small clusters of mature red blood cells (CFU-E), that generate colonies containing multiple erythroblast clusters in culture (mature BFU-E: three to eight clusters, primitive BFU-E: greater than eight clusters). Positive staining with benzidine.

(b) CFU-G. Granulocyte, multishaped nuclei representing the varying stages of differentiation from myelocyte (ring-shaped nuclei in mice) to the characteristic polymorph. Positive myeloperoxidase staining.

(c) CFU-M. Macrophages, round eccentric nucleus, often with cytoplasmic evidence of phagocytosis. Positive with non-specific esterase staining.

(d) CFU-GM. Contains both granulocytes and macrophages.

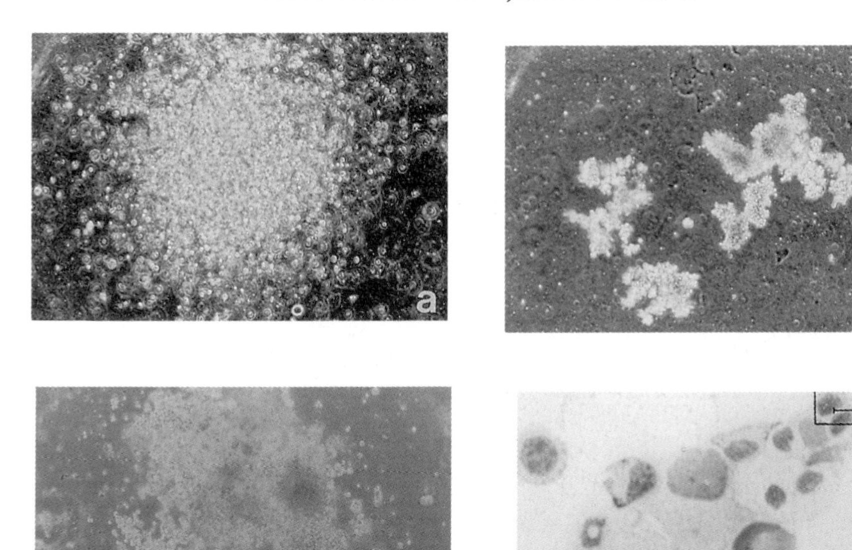

Figure 4. Haemopoietic colony-forming cells. (A) CFU-GM, colony containing granulocytes and macrophages. (B) BFU-E, erythroid burst-forming unit, primitive, more than 8 clusters. (C) CFU-Mix, more primitive colony containing three or more different cell types. (D) Stained (Dif Quik^{tm}) cytospin preparation of CFU-Mix, showing granulocytes, macrophages, and megakaryocytes.

(e) CFU-Mix. More primitive precursor that gives rise to three or more of the following cells types: granulocytes, erythroid cells, macrophages, mast cells, eosinophils, and megakaryocytes. Mast cells contain metachromatic granules that stain positive with Astra Blue, while eosinophils stain bright green with Luxol Fast Blue.

3.3 Myeloid long-term culture for murine and human marrow

The long-term culture (LTC) system, developed first for the murine system (31) and later successfully adapted for human cells (32), appears to mimic the essential cell types and mechanisms responsible for localized stem cell maintenance and continual production of haemopoietic cells. Requisite in this system is the presence of 'stromal' cells that allow very primitive haemopoietic cell survival, differentiation, and proliferation into multilineage precursors in the absence of exogenously added growth factors. The stroma is composed of marrow-derived, mesenchymal cells including fibroblasts, adipocytes, endothelial cells, and smooth muscle cells. It has been shown that these various cell types can be effectively replaced by cloned, immortalized fibroblastoid cell lines of both murine and human origin. This non-specificity suggests that this complex mixture may not be as important as certain common features these cells share, such as cell-mediated interaction and growth factor production.

The LTC system is especially useful in gene transfer analysis of human haemopoietic cells. It provides an *in vitro* assay system for very primitive precursor cells that may be related to the *in vivo* long-term repopulating stem cell. These cells—called long-term culture-initiating cells (LTC-IC)—are distinct from and precursor to most if not all cells detectable in clonogenic assay (33). An arbitrary five week time point as the minimum interval in LTC necessary to detect clonogenic progeny derived from these cells has been established. This time point is based on studies that showed that more mature clonogenic progenitors, present in the initial inoculum, took approximately four weeks to decline to insignificant levels, and that cytotoxic drug-treated cells required at least four weeks to generate significant numbers of clonogenic progeny (34). Additional studies with cells enriched by antibody profiling and adherence panning provide further evidence for the primitive nature of these cells. The determined frequency, by limit dilution analysis of purified fractions, is one per 2×10^4 normal human marrow cells; they produce an average of four colony-forming cells (CFC) per LTC-IC at the five-week time point (35). In the murine system it has been shown that LTC-IC contain long-term myeloid and lymphoid repopulation cells, providing evidence that these cells and the *in vivo* pluripotent stem cell are closely related (36). This relation remains to be proven in the human system.

Protocol 13. Dexter myeloid long-term bone marrow cultures (37)

1. Prepare media for long-term culture (LTC) as follows.
- 73% alpha medium supplemented with 400 mg/litre L-glutamine, 40 mg/litre *i*-inositol (Sigma), 10 mg/litre folic acid
- 12.5% horse serum[a]
- 12.5% FCS[a] (each serum should be used without heat inactivation)
- 10^{-4} M 2-mercaptoethanol
- 10^{-6} M hydrocortisone sodium succinate in alpha medium, added immediately prior to use; do not use after seven days.

2. For both mouse and human cultures seed 3×10^7 BMC in 10 ml of myeloid-LTC media in T25 flask (Costar, 0.2 μm vented filter flask) (*Figure 3*).
(a) Cells are prepared as in *Protocol 9*.
(b) Seed enriched populations and transduced human BM over pre-established stromal feeder layers (see *Protocol 14*). Enriched cells

are seeded at proportionately lower cell concentrations depending on the degree of enrichment expected.

3. Maintain at 37 °C, 5% CO_2 in air for three days, then transfer to 33 °C, 5% CO_2 in air.
4. At seven days, and then weekly intervals, remove half of the media and non-adherent cells, and replenish the cultures with fresh media to 10 ml. The adherent stromal layers will be established by two to three weeks (*Figure 5*).
5. Cultures should maintain active haemopoiesis for at least two to three months, generally longer in mouse cultures. Haemopoietic activity can be monitored by output cell number and CFC assay.

[a] Standardize by pre-testing selected lots of FCS and horse sera. The ability to optimally sustain clonogenic output from a single inoculum of normal marrow for eight weeks in culture is an appropriate end-point.

Protocol 14. Preparation of stromal feeder layers for human LTC

1. Plate cells from Ficoll gradients at 2–3 × 10^7/T25 in 10 ml myeloid-LTC media (hydrocortisone is optional). Incubate at 37 °C, 5% CO_2.
2. Maintain in LTC as in *Protocol 13*.
3. When stromal cells reach confluence (at two to three weeks) trypsinize, resuspend in single cell suspension. Irradiate cells with 15–20 Gy.[a]
4. Replate the irradiated cells at 3 × 10^4 stromal cells/cm^2 or 7.5 × 10^5/T25. Stromal cells are phenotypically larger than haemopoietic cells. Enumerate these cells for plating.
5. Maintain in LTC with weekly half-media change. Use within two to four weeks after plating.

[a] Dependent upon experimental design; autologous layers are not usually irradiated.

3.4 Differentiation of embryonic stem cells to form haemopoietic cells in culture

A new approach to the culture of early haemopoietic precursors may be of interest in gene transfer studies. This approach relies on the capacity of embryonic stem (ES) cells to differentiate into cystic embryoid bodies capable of generating erythroid and myeloid cells in suspension culture (38). The possibility for culture and selection of undifferentiated ES cells followed by differentiation in culture or *in vivo* opens the possibility of genetic studies that would be difficult if not impossible in transgenic mice or conventional marrow

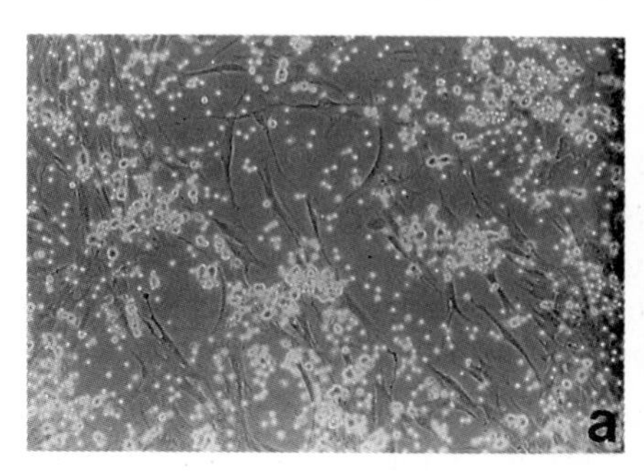

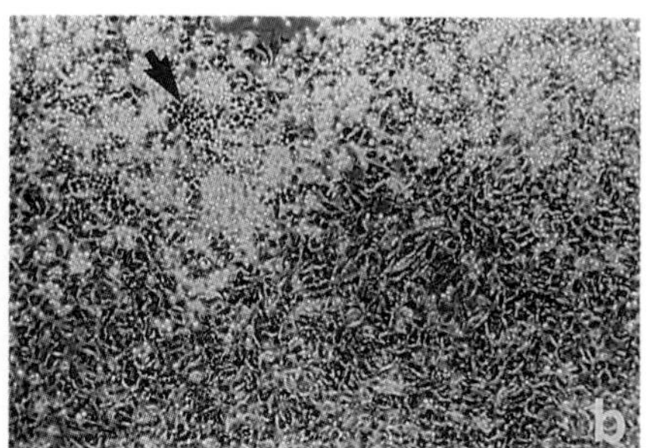

Figure 5. Long-term bone marrow cultures. (A) Developing bone marrow stromal layer five to seven days after initial seeding. (B) Established stromal layer at four weeks of culture. *Arrow* indicates 'cobblestone' area of active haemopoiesis.

cultures. The combination of ES cell culture techniques with gene transfer and subsequent generation of whole animal chimeras should prove very powerful in evaluating the regulation of gene expression, dominant or hemizygous mutations, vector expression, and cell interactions. Various growth factor combinations have been described in preliminary reports of this assay system. These include addition of IL-1, SCF (stem cell factor, c-kit ligand), LIF (leukaemia inhibitory factor), IL-7, and IL-4 to these cultures (39). It is expected that growth factor combination, dose, timing, and order of exposure will all affect the ultimate production of mature cells. The goals of each experiment will determine the optimal use of growth factors.

Protocol 15. Embryonic stem cell differentiation

1. Maintain AB1 ES cells (A. Bradley, Houston, TX) (40) in maintenance media:
   - DMEM
   - 15% FCS (tested lot)
   - 150 μM monothioglycerol (MTG, Sigma, tissue culture grade)
   - LIF (either recombinant murine LIF 1000 U/ml or STO feeder layer)
2. Transfer cells to IMDM-based medium with identical supplements one to five passages before differentiation.
3. Harvest cells at 50% confluence in 0.25% trypsin, 1 mM EDTA. Generate a single cell suspension.
4. Transfer 1–2 × 10^3 single cells to 1.25 ml differentiation medium in a 35 mm suspension culture dish. Embryoid bodies should begin to form after three days, expect 50–100 to form.

   Differentiation medium
   - 0.9% methylcellulose, final concentration from stock (see *Protocol 12*)

- 10% FCS
- 450 μM MTG

5. Add supplements to maintain erythropoiesis and myelopoiesis.
   (a) Erythropoietin (2 U/ml) at nine days to maintain erythropoiesis in 40–60% of the embyroid bodies. Erythropoiesis begins after 7–8 days, reaches a peak at 9–11 days, and will decline unless erythropoietin is added.
   (b) Vitamin E (50 μg/ml) or ascorbic acid (50 μg/ml) will increase macrophage frequency to 10–15% from 5–10% after 12–18 days when added initially.

---

4. *In vivo* models for murine haemopoiesis

4.1 Proliferative and developmental potential of murine stem cells

In vivo, haemopoietic stem cells proliferate and develop in association with bone marrow stromal cells. In the bone marrow, the stromal cells and mature haemopoietic cells produce growth factors which create micro-environments for lineage determination, proliferation, and self-renewal. Although culture assays have proven useful in the analysis of committed progenitors, it has been difficult to correlate the growth of cells in culture with the most primitive cells of the developmental hierarchy. Analysis of these cells has depended upon bone marrow transplantation into lethally irradiated recipients. The most accessible of these primitive cells are the myeloerythroid-restricted precursors that home to the spleen of recipient animals after injection. These cells—termed colony-forming unit, spleen (CFU-S)—were first described in 1961 by Till and McCulloch (41), who demonstrated a linear relationship between splenic nodules or 'colonies' present after 8–12 days and the number of cells injected (*Figure 6*). They also hypothesized and later confirmed that these were clonal in origin, that is, the colony developed from a single cell. For a number of years this was considered to be an assay for a pluripotent haemopoietic stem cell. Subsequent evidence based on kinetics of colony formation, retransplantation, clonal analysis with genetic markers, and differential purification indicates that these cells are not active in long-term repopulation. More recently the concept of the marrow repopulating CFU-S day 12 (MRA-CFU-S-12) has been introduced and used to characterize a pre-CFU-S cell (50). These cells are present in the marrow 13 days after initial transplant and are defined by their ability to generate CFU-S-12 in a secondary recipient. The relationship between the MRA-CFU-S-12 and the long-term repopulating stem cell is an area of active investigation. These types of *in vivo* assays have broad application in retroviral vector-mediated gene transfer studies. They have been used extensively to identify conditions that enhance stem cell transduction by expanding or causing proliferation of these various stem cell pools (43, 44).

To study reconstitution in murine transplantation models various properties unique to the repopulating cells have been exploited. In addition to using a Y chromosome marker to study repopulation in sex-mismatched transplants, various polymorphic loci on congenic backgrounds have proven useful for analysing the activity of repopulating cells in mature haemopoietic lineages. These include but are not limited to: haemoglobin, glucose-phosphate-isomerase, Thy-1, and Ly-5. These markers have been particularly informative in the analysis and development of stem cell enrichment pro-

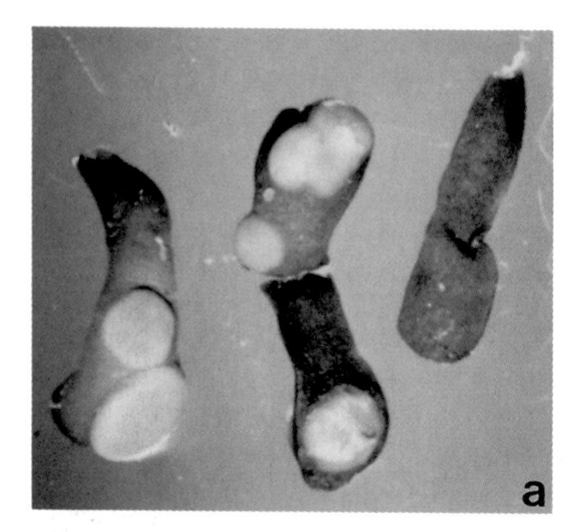

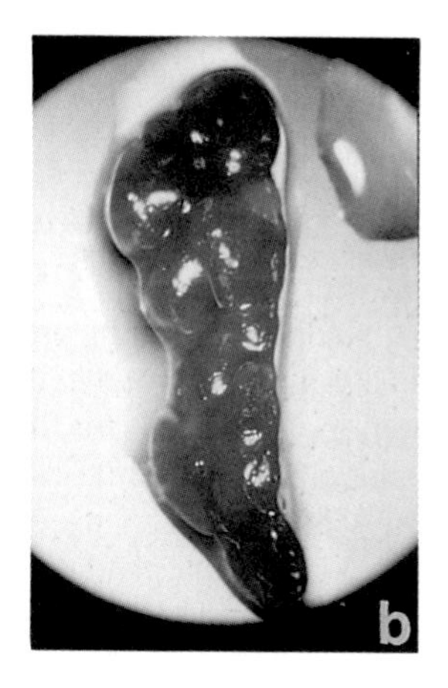

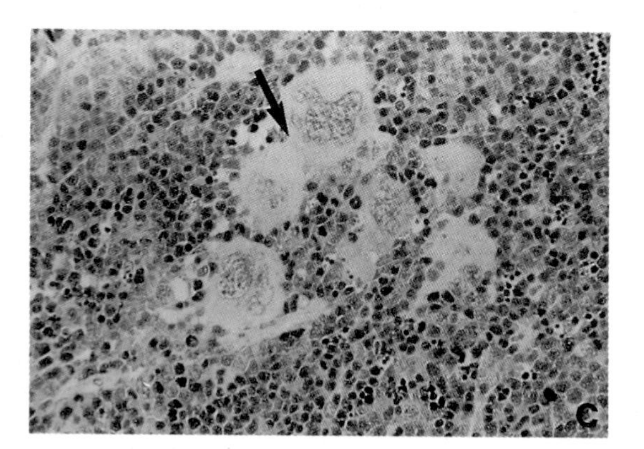

Figure 6. Mouse spleen colony-forming units, CFU-S. (A) Bouin's staining of spleens, discrete, white colonies on two spleens, the third spleen is from a non-transplanted, radiation control, showing no appearance of endogenous colonies. (B) Fresh, non-fixed spleen collected 12 days after transplant. (C) Spleen section (H&E stain) depicting a mixed-lineage CFU-S consisting of primarily megakaryocytes, indicated by *arrow*.

tocols. CFU-S-12 assays have been used to determine the relative amount of enrichment by limit dilution analysis. In some protocols, it can be shown that CFU-S-12 are physically separable from MRA-CFU-S-12 (42). Highly purified marked cells have been used in competitive repopulation assays with stem cell depleted populations to prove the proliferative and pluripotent nature of the purified cells. In comparison, stem cell marking by retroviral vector insertion offers the unique advantage of allowing determination of clonal relatedness among groups of repopulating cells. We have recently combined these approaches by using a competitive repopulation assay with neutral transgenic markers and retroviral insertion to evaluate methods for gene transfer of human *adenosine deaminase* (*Figure 7*) (45).

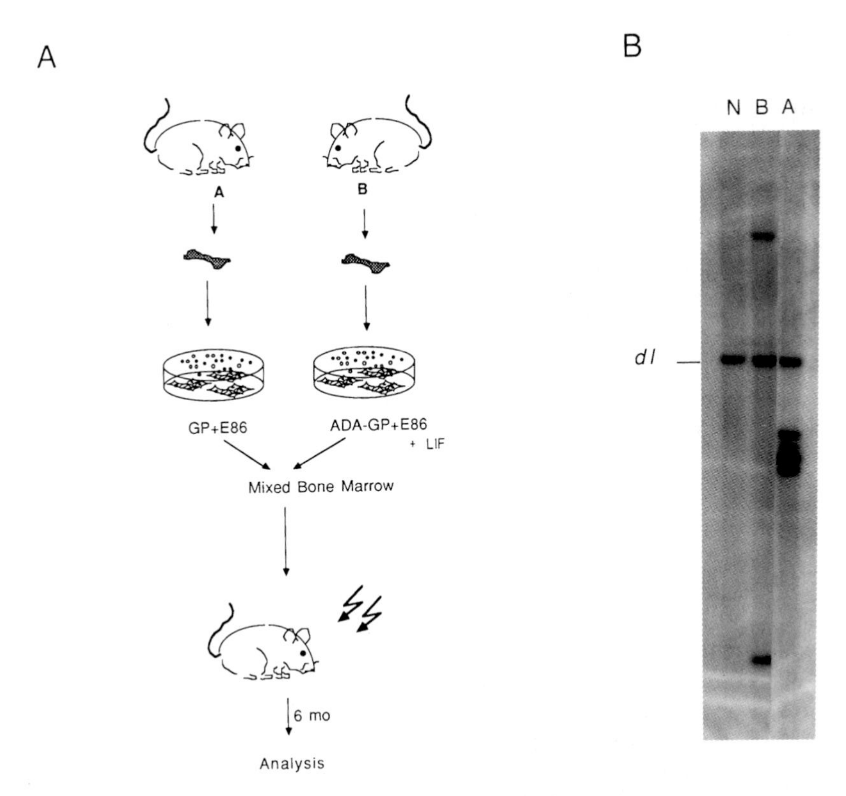

Figure 7. Competitive-repopulation assay. (A) Bone marrow cells are collected from two transgenic lines, carrying the same neutral marker but in different chromosomal locations and copy number. These cells are then treated under differential conditions *in vitro* (one retroviral vector-transduced the other not), combined, transplanted in a lethally irradiated non-transgenic, syngeneic recipient. At six months the haemopoietic tissues of the repopulated mice are analysed for both the presence of the transgenic and retroviral markers. (B) Southern blot analysis of DNA from non-transgenic mouse (N), and both transgenic lines studied, (A and B). *dl*, downless locus; used as control for DNA loading. (Reproduced from *J. Exp. Med.*, 1991, **174,** 837.)

4.2 Apparatus

(a) Microisolator system (Lab Products Inc., Maywood, NJ).
  i. Microisolator cages.
  ii. Laminar flow cage changing hood.
  iii. Laminar flow cage racks.

(b) Small animal irradiator. Many investigators prefer the Gammacell 40 ^{137}Cs source because of dose-rate, and field homogeneity (Nordion, Ontario, Canada).

(c) Dissecting microscope.

4.3 General concerns about housing and maintenance of specific pathogen free (SPF) mice

Mice can be purchased SPF from reliable vendors. Recent screening information should be enclosed with each shipment or be available on request. Our transplantation models have been conducted with the FVB/N mouse strain. The transgenic lines were generated in this strain by P. Overbeek (Department Cell Biology, Baylor College of Medicine, Houston, TX), further Caesarian-derived, and maintained SPF in our colony. Mice can be maintained SPF in a monitored barrier facility or in a conventional facility with adequate precautions. Mice to be rendered immunodeficient by lethal total body irradiation (TBI) should be housed in autoclaved microisolator cages with autoclaved bedding, food, and water. Hepa-filtered ventilated cage racks and special changing hoods provide a necessary, additional level of safety in a conventional facility with endemic mouse viruses. The colony should be monitored quarterly for mouse viral pathogens (Mouse level II complete antibody profile, Microbiological Associates Inc., Rockville, MD), bacterial pathogens, endo- and ectoparasites.

4.4 Irradiation of recipient animals

The total dose is dependent on the strain of mice used and should be characterized for each irradiator and strain. The usual dose is 8.0–12 Gy. Lower dose-rates (0.3–1.0 Gy/min) are preferred. Non-transplanted radiation control mice should live for 14–21 days if housing and radiation standards are adequate. The mice should be transplanted within 24 hours of lethal TBI.

Protocol 16. Primary CFU-S, day 12 (CFU-S-12)

1. Inject cells intravenously into the retro-orbital venous plexus with a blunt needle or by lateral tail vein (27 gauge needle). Anaesthesia is required for retro-orbital injection. Use a 50 ml conical tube containing a gauze dampened with methoxyflurane (Metofane^{tm}) for the anaesthetic chamber. Monitor anaesthesia by toe pinch, (withdrawal of limb) and respiration (rate and depth). Mice recover in less than 1–2 min.

2. After 12 days collect the spleens.
   (a) Fix in Bouin's solution (*Figure 6a*), consisting of:
      - saturated picric acid 75 ml
      - 37% formalin 25 ml
      - concentrated acetic acid 5 ml

      and enumerate colonies with the aid of a dissecting microscope.
   (b) If gene transfer efficiency is to be assessed; carefully dissect large, discrete colonies (approximately 10^6 cells) without fixation (*Figure 6b*). Whole marrow recovered after retroviral transduction will yield approximately 10–20 CFU-S-12 per 10^5 cells transplanted. Cells can be used for:
      - protein extracts for expression studies of the transferred gene
      - DNA for provirus analysis
      - CFU-S assays.
   (c) To identify microscopic colonies and determine the cellular composition of all colonies, fix spleens in neutral buffered formalin, section, and H&E stain. (*Figure 6c*).

Protocol 17. Marrow repopulating ability-CFU-S, day 12 (MRA-CFU-S-12)

1. Transplant lethally irradiated mice with 1×10^6 whole marrow cells. If using enriched cell populations, cell dose is dependent on the purification scheme. Quantitation of MRA can be approached by limiting dilution of the marrow inoculum in the primary recipients.
2. After 13 days obtain the marrow from one femur, see *Protocol 9*.
3. Perform viable cell count and inject secondary lethally irradiated recipients with 1×10^5 cells.
4. Maintain 12 days, then harvest spleens, and count colonies or dissect as above.

Protocol 18. Long-term reconstitution assay—competitive repopulation

1. Transplant lethally irradiated mice with 1×10^7 marrow cells.
   (a) For standard competitive repopulation studies transplant equal number of differentially treated cells.
   (b) Quantitation of repopulating cells can be further enhanced by limiting dilution of one marked population while using the second as 'filler' to maintain a constant cell dose.

Protocol 18. *Continued*

2. Allow mice to reconstitute for at least six months.
3. Engraftment can be studied by measuring the percentage of mature cells in various lineages bearing the genetic marker. Alternatively, colony assays and/or CFU-S assays can be used starting with marrow cells of the primary recipients. This is particularly helpful since it allows precise quantitation within the precursor pools.

Protocol 19. Retroviral vector transduction of murine HSC

1. Prepare bone marrow/retrovirus transduction culture medium (BMM).
   - IMDM, base media
   - 10% BCS
   - 10 mg/ml BSA (Boehringer Mannheim, tissue culture grade)
   - 0.0017% α-thioglycerol (Sigma, tissue culture grade)
   - 4 μg/ml polybrene
   - 100 U/ml penicillin–100 μg/ml streptomycin (Gibco)
   - 10 U/ml recombinant murine LIF (commercial source)
2. Establish a monolayer of irradiated virus-producing cells.
   (a) Trypsinize virus-producing cells from logarithmic growth phase culture (~70% confluent flasks). Wash cells, and resuspend in growth media at 1–5 × 10^6/ml.
   (b) Irradiate (20 Gy) cells in suspension.
   (c) Plate irradiated cells at 3 × 10^6 cells in 100 mm tissue culture dishes. Allow cells to adhere at least 4 h prior to addition of BMC (obtained as described in *Protocol 9*).
3. Remove media and add 3 × 10^6 BMC in 10 ml BMM to established virus-producing monolayers. Co-cultivate for 72 h at 37°C, 5% CO_2.
4. Harvest marrow cells from dishes.
   (a) Remove non-adherent cells and save.
   (b) With fresh media vigorously wash the fibroblast layers to dislodge adherent marrow cells.
   (c) Allow fibroblast clumps to settle by gravity for 10 min.
   (d) Repeat gravity sedimentation if necessary to remove clumps.
   (e) Combine fractions, wash twice in IMDM, prepare single cell suspension, and perform viable cell count.

5. Inject transduced marrow cells IV into lethally irradiated syngeneic recipients.
   (a) Cell number injected is dependent on assay conducted.
   (b) For long-term transplant controls include:
      i. Mice injected with the mock-infected marrow (cells incubated over packaging cells without vector).
      ii. Radiation control, media only injection.
6. Although transduction is vector-dependent you may expect transduction efficiencies of 50–100% under these conditions, using a vector with a titre of $\geqslant 10^6$ units/ml.

4.5 Enrichment of murine haemopoietic stem cells

Several recent reviews provide excellent summaries of different approaches to HSC purification (46, 47). These include density centrifugation, counterflow centrifugal elutriation, monoclonal antibody staining, and multiparameter-fluorescence activated cell sorting (FACS). Use of monoclonal antibodies and FACS has led to isolation of murine stem cell populations highly enriched for CFU-S and LTR cells. However further separation of these cells on the basis of other cell surface markers (c-kit) and supravital staining (Rh 123, H33342) has indicated that different subpopulations of stem cells have to be distinguished that have different capacities regarding self-renewal, *in vitro* growth and differentiation, radioprotection, and long-term repopulation of haemopoiesis. A reproducible procedure for the ultimate purification of pluripotent stem cells will bear important biological and medical implications. It should allow molecular characterization of the complex mechanisms (growth factor receptors, adhesion molecules) that regulate growth and differentiation of HSCs. Their purification will also facilitate development of an *in vitro* culture system to maintain them in a pluripotent state, and development of protocols for efficient transfer and expression of foreign genes in stem cells; technologies that could play an important role in the treatment of some genetic and acquired diseases. The basic techniques provide a starting point from which further enrichment procedures may be attempted. We include two protocols both of which give approximately 100–200-fold enrichment of CFU-S. Whether absolute purification of stem cells has been achieved by any group is a matter of debate. We prefer the term 'enrichment' to reflect the uncertainty about purity of separated cell populations and the bioassays for their activity. A crucial difference between our procedures and several of those used by other groups is the pre-treatment of the source animals with 5-flourouracil. This results in an approximately three-fold enrichment for CFU-S via reduction of drug-sensitive mature cells. It almost certainly also causes *in vivo* stimulation of the precursor pools which may lead to changes in phenotype, responses to cytokines, and bioactivity in repopulation assays. The goal of the individual experiment, therefore, should determine whether this is an appropriate manoeuvre.

4.5.1 General considerations

(a) Preparation of the metrizamide density gradient solutions.

i. H.HBSS—Hepes–Hanks' buffered salt solution—add 5.66 g Hepes to 2 litres HBSS, pH 6.9, 300 mOsm.

ii. Low density (1.055 g/cm^3, pH 6.9, 300 mOsm)—mix 11 g metrizamide, 1 g BSA, 64 ml H.HBSS (pH 6.9, 300 mOsm). Add water to 100 ml.

iii. High density (1.1 g/cm^3, pH 6.9, 300 mOsm)—mix 21 g metrizamide, 1 g BSA, 42 ml H.HBSS. Add water to 100 ml.

iv. Intermediate density solution (1.072 g/cm^3, pH 6.9, 300 mOsm) is prepared by mixing high and low density solutions (37.78 ml high plus 62.22 ml low for each 100 ml of intermediate).

(b) Density, osmolarity, temperature, and centrifuge configuration are critical variables in the performance of the gradient. Density of the final solution at 7°C should be measured using an accurate temperature controlled density meter (Mettler DA-300, Mettler Instrument Corp., Highstown, NJ). Osmolarity and pH should also be measured.

(c) An alternative to metrizamide is Percoll.

i. Prepare Percoll solutions with densities of 1.10, 1.09, 1.07, and 1.06 g/cm^3 according to the protocol recommended by the manufacturer (Pharmacia Fine Chemicals).

ii. Adjust the pH of solutions to pH 7.0 and osmolarity to 300 +/− 5 mOsm/kg. Form the gradient by layering sequentially 2 ml of 1.10, 1.09, and 1.07 g/cm^3 solution.

iii. On top of the gradient lay 2 ml of 1.06 g/cm^3 solution containing up to 5×10^7 cells.

iv. Keep all solutions sterile and at 4°C.

(d) Monoclonal antibodies. Lineage-specific cell surface markers can be used for identification and depletion of mature cells. Monoclonal antibodies specific for murine CD4 (L3T4, GK1.5), CD8 (Lyt2, 53-6.72), Mac-1 (M1/70), and B220 (RA3-6B2) are widely available from commercial sources. Monoclonal anti-Ly6.2A (Sca-1, E13-161-7) provides a marker for HSC in the lineage negative, Thy-1^{lo} subpopulation (49). Monoclonal anti-mouse monocyte–macrophage (G15.1.1) is useful for depletion of monocytes and degranulated PMN's which have density and FACS profiles similar to HSC (48). Inquiries for these antibodies should be made to Dr I. L. Weissman for Sca-1 (Stanford University, Palo Alto, CA) and Dr J. W. M. Visser for G15.1.1 (Radiobiological Institute TNO, Rijswijk, The Netherlands).

Protocol 20. Enrichment of murine haemopoietic stem cells

Method A

1. Remove marrow cells from 6–12 week old mice (see *Protocol 9*). Dissection may be performed the evening before and intact bones stored in IMDM at 4°C overnight.
2. After dispersal into single cell suspension determine viable cell count.
3. Centrifuge cells at 500 *g* and resuspend at 1×10^8/ml. For each 100 μl add 100 μl wheat germ agglutinin-FITC conjugate (80 μg/ml) (Polysciences, Warrington, PA). Mix thoroughly and protect from light.
4. Load $1–3 \times 10^7$ viable nucleated cells on top of a three layer discontinuous metrizamide gradient in round-bottom polystyrene tube; bottom—high density 1.0 ml, middle—intermediate density 3.5 ml, top—low density 2.0 ml. Centrifuge at 1000 *g*, 7°C for 20 min.
5. Gently harvest cells at the low:intermediate density interface and wash twice in H.HBSS. Count cells—the yield should be approximately 5–7%. The morphology of the cells on Giemsa staining shows a variety of blasts, lymphocytes, monocytes, and band cells.
6. Immunomagnetic depletion for mature lineage markers may be performed at this point. Alternatively, the mature lineage markers may be used for counter-sorting by FACS. The G15.1.1 marker has been shown to be useful in either mode.
7. Resuspend the cells in H.HBSS at 1×10^6/ml for FACS. Gate the FACS for the medium forward light scatter, low side scatter, and very high WGA-FITC fluorescence (upper 50% of positives) (*Figure 8*). The overall yield of cells from these steps is approximately 0.5% of starting cell number, with a relative enrichment of CFU-S of approximately 100-fold. This population also contains MRA activity.
8. Second sorting may be used for further enrichment, e.g. the use of rhodamine-123 exclusion (48) and Sca-1 staining.

Method B

1. Prepare bone marrow cells from ten mice as above.
2. Resuspend up to 5×10^7 cells in 2 ml of low density (1.06 g/cm^3) Percoll solution and load on top of a three-layer discontinuous Percoll gradient in round-bottom polystyrene tubes. Centrifuge at 1000 *g*, 4°C for 25 min.
3. Harvest the low density cells at the 1.06/1.07 g/cm^3 interface and wash twice in cold 0.15 M NaCl. Count the viable cells (yield $\sim 4 \times 10^7$, i.e. 13%) and resuspend in PBS with 5% FCS and antibiotics (10^7 cells/ml).

Protocol 20. *Continued*

4. Incubate the cells with saturating concentrations of a cocktail of lineage-specific rat monoclonal antibodies, directly conjugated with FITC. After 35 min incubation on ice wash the cells twice (800 *g* for 10 min) and resuspend at the same density.
5. Incubate the cells with rat anti-Ly6.2A (Sca-1) for 35 min on ice. Wash the cells twice with PBS, incubate for another 30 min with phycoerythrin (PE)-conjugated $F(ab')_2$ goat anti-rat IgG (Jackson ImmunoResearch), and repeat the wash. Count the viable cells and resuspend at 3×10^6/ml in PBS with 5% FCS and antibiotics.
6. Analyse the cells on a pre-sterilized dual laser FACStar-Plus instrument (Becton Dickinson) and set the gates for four-parameter sorting. Select the cells for sorting on the basis of intermediate forward scatter (to exclude small lymphocytes and debris), low side scatter (to exclude highly granular cells), low levels of FITC fluorescence (lineage negative cells, Lin^-), and high levels of PE fluorescence ($Sca\text{-}1^+$ cells). Keep the cells at 4°C.
7. Sort the cells at a rate of 1000–2000 cells per second and collect them in cold PBS containing 10% FCS and antibiotics. After completion of the sort (yield is approximately 0.05–0.1% of starting cell number; the purity and viability of sorted cells should be about 90% or more), wash the cells twice (800 *g* for 10 min), and resuspend as required for assays.

5. Human haemopoietic cell analysis

5.1 Mouse/human chimeras as assays for human gene transfer

Immune deficient mouse strains have been recently described as *in vivo* models for human haemopoiesis. These models provide a new approach for characterization of human stem cells and offer hope as animal models for many diseases. They are the severe combined immune deficiency *scid/scid* (scid) and *beige/nude/xid* (bnx). The scid mutation arose spontaneously in the C.B.-17 inbred strain (51). These mice are severely deficient in functional T and B lymphocytes. The mutation appears to impair the recombination of antigen receptor genes and is reflected in other cells as a general defect in DNA repair. The triple mutation bnx mice were developed in an attempt to generate a better model for human tumour xenografts. The *beige* mutation results in a deficiency in cytotoxic T cells and NK cells due to a lysosomal storage defect. The *nude* mutation affects the development of the thymic epithelium, which limits T cell differentiation and results in an athymic mouse. The *xid* defect affects the B cell response to some thymus-independent antigens and lymphokine-activated killer (LAK) cells.

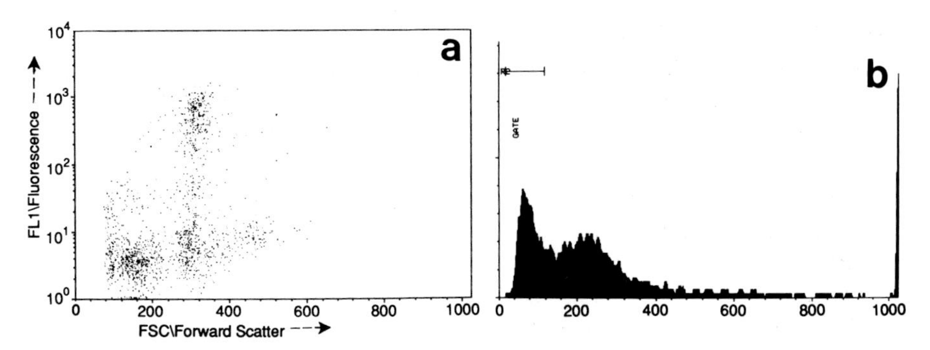

Figure 8. FACS profiles of haemopoietic progenitor cells. (A) Scatter plot showing forward light scatter and fluorescence with WGA-FITC of low density bone marrow cells. (B) Right angle light scatter of low density cells from scatter plot. The low right angle light scatter population is shown by the *gating arrow*. The WGA-positive, medium forward light scatter and low right angle light scatter population represents approximately 6–10% of the total cells profiled. (FACS analysis by Dr N. T. Van: Facstar Plus, Becton Dickinson, Argon Laser; 488 nm excitation, 520 nm band pass.)

Different approaches to using these mice for demonstration of *in vivo* haemopoiesis and human gene transfer have been developed. A summary of these strategies was recently reviewed by Dick *et al.* (52). Production of human immunoglobulin and lymphomogenesis has been demonstrated following intraperitoneal injection of peripheral blood leucocytes to generate scid-PBL. Others have pursued an approach based on the importance of an appropriate micro-environment for human haemopoietic cell survival. Chimeric (SCID-hu) mice were constructed with grafts of human fetal tissue, in particular liver and thymus grafts under the kidney capsule, and reconstituted intravenously with fetal liver cells. In attempts to promote bone marrow engraftment, others have used sub-lethal irradiation prior to intravenous injection. It is believed that sub-lethal irradiation impairs residual activity of immunoactive cells and stimulates production of cross-reacting cytokines. Our laboratory has used both strains in human bone marrow gene transfer assays and the generation of scid-PBL. The bnx model has proven to be more difficult to maintain and breed and will not be discussed further. Scid mice breed well and have proven a more useful model in our hands. The following is offered as a guideline only, as these protocols are still in the developmental stage. We have not generated SCID/hu mice and they will not be described further. Scid-PBL mice will not be described as to limit this discussion to haemopoietic stem cells.

5.1.1 General considerations

(a) Sub-clinical infection will lead to graft rejection due to hyperactivity of NK cells. Constant low level pathogens in the environment will increase the percentage of 'leaky' mice.

(b) 'Leaky scids' are mice that gain partial immune competence via reversion of the mutation in somatic clones. Approximately 15% of young adult scid mice are 'leaky' and virtually all are beyond one year of age. Therefore, monitor the colony for mouse immunoglobulin production and conduct experiments with mice less than six months old.

Protocol 21. Source, housing, and maintenance of scid mice

1. Obtain scid mice from commercial or academic source, e.g. the Ontario Cancer Institute, Toronto, Canada.
2. Maintain mice under extremely rigid, clean environmental standards, isolated from the rest of colony.
   (a) Maintain separate supplies and sterile food.
   (b) House mice in autoclaved microisolator cages on a hepa-filtered ventilated cage rack (Laboratory Products, Inc.).
   (c) Conduct all manipulations in a laminar flow changing hood. Cover all skin surfaces, and periodically disinfect gloves and all surfaces with spray sterilant, e.g. Alcide^{tm} (Fisher).
3. Monitor mice as follows.
   (a) Food, water, and dirty bedding should be routinely monitored for adventitious bacteria. Mice should be able to survive 4.0 Gy irradiation. As experimental controls irradiated mice provide a useful screen for the presence of sub-clinical infections in the colony.
   (b) Assay for production of mouse immunoglobulin. This is most easily accomplished by ELISA using commercially available reagents (Sigma).

Protocol 22. Transplantation of human marrow cells into scid mice

1. Give recipient scid mice 4.0 Gy TBI.
2. Inject at least 2–4 × 10^7 human BMC IV. Begin analysis one to two weeks after transplant.
3. On alternate days following transplantation inject IP first 20 μg stem cell factor (SCF, MGF, kit ligand, can be obtained in limited quantities from Immunex or Amgen), and than 20 U EPO (Amgen). This treatment is continued throughout the course of the experiment. Recently Lapidot *et al.* (53) have described this modification of the original method that greatly enhances the growth and recovery of human cells from transplanted scid mice. The amount of growth factor used is large, however, and this may limit the application.
4. Assay for human IgG by ELISA using commercially available reagents (Sigma).

5. Use flow cytometry to assess for human cells in peripheral blood, peritoneal cells, spleen, and bone marrow. Directly conjugated anti-CD45 is useful since it is expressed on all lineages.

6. Prepare samples of DNA from fractions of spleen, bone marrow, and peritoneal cells for PCR. For 100–200 μl whole blood total nucleated cell count should be $<10^6$.

(a) Collect two heparinized microhaemocrit tubes per mouse by retro-orbital bleed (one tube holds 75 μl). Expel into 1.5 ml microfuge tube. Centrifuge, save plasma for ELISA.

(b) To cell fraction add 500 μl cold, red cell lysis buffer (*Table 2*). Mix well. Centrifuge 13 000 *g* for 30 sec. Remove supernatant, (individually with clean pipette tip). Repeat one to three times until haemoglobin removed.

(c) To cellular DNA pellet add 100 μl quick lysis buffer (*Table 2*), mix well, incubate overnight at 37°C or 4 h at 55°C. Heat 85–90°C, 10–15 min to inactivate proteinase K. Centrifuge to pellet cellular debris. Use 3–5 μl/50 μl PCR reaction.

7. Carry out PCR to detect human specific sequences.

(a) Oligonucleotide primers: (R. C. Allen, Houston, TX)
sense 5′[CTGGGCGACAGAACGAGATTCTAT]3′
antisense 5′[CTCACTACTTGGGTGACAGGTTCA]3′

(b) Reaction mixture for human-specific PCR

- 1 × PCR buffer (*Table 2*)
- 0.2 μM each primer
- 0.25 mM each dNTP
- DNA 3–5 μl/50 μl reaction (from quick prep)
- water to volume

(c) Denature 5 min 95°C. Cycle 35–40 cycles: 95°C 30 sec, 60°C 30 sec, 72°C 30 sec.

(d) Electrophorese 15–25 μl reaction mixture through 1% agarose + 1% Nusieve gel. Visualize 200 base pair band by ethidium bromide staining. A starting template of 0.1 ng human DNA will yield an amplification product visible by ethidium bromide staining.

8. Perform human-specific colony assay.

(a) Assay mixture (final concentration):

- MC (*Protocol 9*) (0.8%)
- human plasma (pre-tested) (30%)
- 2-mercaptoethanol (5×10^{-5} M)
- recombinant human IL-3 (1 U/ml)

Protocol 22. *Continued*

- recombinant human GM-CSF (9 U/ml)
- 2×10^5 cells/ml in IMDM

(b) Harvest cells from each mouse (*Protocol 9*) and separate over standard Ficoll gradient. Count the nucleated cells from the light density fraction and plate.

(c) Plate and maintain as in regular colony assay (*Protocol 12*). Colonies may be picked and assayed by human-specific PCR (see above) for confirmation, and by other PCR assays if assessing gene transfer.

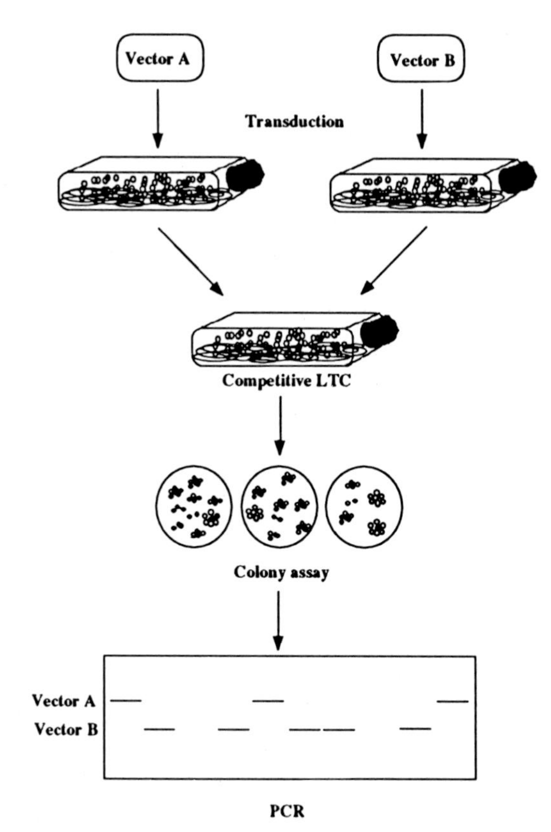

Figure 9. Competitive long-term bone marrow culture. Two vectors that are distinguishable in a single PCR reaction are used to transduce two differentially treated populations of BMC. Differential conditions may include: comparison of growth factor stimulus, evaluation of purging protocols for leukaemic marrows, and evaluation of enrichment protocols. The cells are transduced by cell-free virus-containing supernatant over pre-established bone marrow stromal layers, combined, and maintained in LTC. Clonogenic progenitors from LTC-IC are grown in colony assay, individual colonies are picked and analysed by PCR for the presence of the transducing vector.

5.2 Transduction of human haemopoietic cells

Considerable progress has been made towards gene transfer into human bone marrow cells. Previously, successful gene transfer into human BM LTC-IC had only been achieved in studies where transduction was accomplished by co-cultivation of marrow cells with virus-producing cells. For a review see Karlsson (54). Supernatant or cell-free transduction of LTC-IC has been recently demonstrated (55). This protocol capitalizes on the supportive micro-environment that is provided by a pre-established bone marrow stromal layer as the context in which to transduce the experimental human marrow. In considering possible application to human gene therapy, supernatant transduction is desirable for reasons of biosafety and quality control. The cell-free transduction protocol has also been further refined to include the option of establishing competitive LTC (*Figure 9*). This system allows analysis of LTC-IC derived from clonogenic progenitors that have been differentially manipulated and then maintained in one micro-environment. Two vectors distinguishable by a single PCR reaction are used to transduce these cells separately and then they are combined for LTC. A positive effect of an experimental manipulation during transduction, for example, a growth factor effect, would be reflected by the presence of a predominant provirus in the PCR assay. These assays may also have application in analysis of leukaemic marrows, such as in evaluating purging techniques, by comparing differentially marked cell populations to determine the source population of leukaemic clone recurrence. This section will outline both procedures that have been successful, in our hands, for transduction of human bone marrow LTC-IC.

Protocol 23. Co-cultivation (56) for transduction of human haemopoietic cells

1. Irradiate amphotropic virus-producing fibroblasts in logarithmic growth phase (20 Gy).
2. Plate the fibroblasts in a 10 cm tissue culture dish, 3–5 × 10^6 cells/dish, 12 h prior to seeding with bone marrow, in 10 ml growth media.
3. Isolate low density (d < 1.077) BMC over standard Ficoll gradients. Wash twice with D-PBS. Perform viable cell count.
4. Add BMC, 3–5 × 10^6/dish of virus-producing cells in 10 ml IMDM supplemented with 12.5% FCS, 12.5% horse serum, 4 μg/ml polybrene, recombinant human IL-3 10 U/ml, recombinant human IL-6 200 U/ml. Recombinant growth factors are available from commercial sources.
5. Incubate cells at 37°C, 10% CO_2 for 48–60 h.
6. Collect non-adherent cells, then detach adherent BMC by vigorous washing with D-PBS. Filter through sterilized 20 μm nylon mesh to remove clumps of virus-producing cells.

Protocol 23. *Continued*

7. Perform viable cell count, then plate 5–10 × 10^6 BMC over an irradiated pre-established stromal layer in a T25 flask, and maintain in LTC (see *Protocol 13*).
8. Sample cultures for transduction efficiency and gene expression at regular intervals (see below).

Protocol 24. Cell-free transduction of human haemopoietic cells

1. Harvest virus in batches from large cultures (T150 or T225 flasks) 24 h after placing fresh media over subconfluent (~80%) producer cells.
   (a) Filter (0.45 μm) the culture supernatants and aliquot.
   (b) Freeze viral stocks at −70°C. Titre should be determined with each experiment.
   (c) The media can be either IMDM with 10 mg/ml BSA, 0.0013% α-thioglycerol, 20% BCS, or the above co-cultivation media.
2. Transduce BMC as follows.
   (a) Prepare BMC as above or as preferred for experimental application.
   (b) Resuspend 1 × 10^7 cells in 5 ml of freshly thawed virus-containing supernatant (quick thaw at 37°C). Add polybrene to 4 μg/ml.
   (c) Place over pre-established irradiated allogeneic or autologous stromal layers, and incubate at 37°C, 5% CO_2 in air for 12 h.
   (d) After 12 h add 5 ml of fresh virus and polybrene, incubate 12 more hours.
   (e) Remove media containing BMC, pellet cells, and resuspend in 5 ml of fresh virus and polybrene. Incubate 12 h.
   (f) Repeat as above for a total incubation time of 72 h.
3. Transfer transduced cells to LTC.[a]
   (a) Collect culture supernatant with non-adherent cells. Vigorously wash the stromal layers with 2% FCS in D-PBS, combine.
   (b) Treat the stromal layers with collagenase to detach adherent haemopoietic cells.
      i. Prepare collagenase (200 U/mg protein, Sigma) at 0.1% in IMDM with 20% FCS, filter sterilize.
      ii. Add 3 ml to each T25 flask and incubate for 3 h at 37°C.
   (c) Wash layers with 2% FCS in D-PBS by sustained gentle pipetting. This procedure routinely leaves 25–30% of the fibroblastic stromal cells in the flasks which do not contain haemopoietic progenitors.

(d) Combine all fractions and wash twice with 2% FCS in PBS.

(e) Transfer 5×10^6 cells to fresh stromal layers.

4. Maintain transduced cells in LTC (*Protocol 13*).

[a] If transducing autologous BMC maintain in same LTC flask. These feeder layers are generally not irradiated.

Protocol 25. Controls for measurement of transduction in human haemopoietic cells

1. Seed non-manipulated cells directly into LTC.

2. Manipulate cells without virus. For co-cultivation, place BMC over non-transfected packaging cells, e.g. AM12. For cell-free, 'mock' transduce cells in media over stroma.

3. Transduce stromal layers with virus in absence of BMC. Use these stromal layers for LTC seeded with non-manipulated non-infected BMC.

4. Maintain the above in LTC and assay alongside experimental samples for progenitor cells, harvest colonies, prepare DNA as above, and use for negative controls in PCR assays.

Protocol 26. Gene transfer analysis for human haemopoietic stem cells

1. Assess transduction efficiency by plating cells into colony assay (*Protocol 12*) at one to two week intervals, and analyse DNA from cells by PCR.

2. Pick well isolated individual colonies under microscopic guidance, and expel into 250 μl of D-PBS in 1.5 ml microcentrifuge tube on ice.

(a) An apparatus composed of a mouth piece with tubing attached to a 0.22 μm syringe filter also attached to a short piece of tubing that will hold a small pipette tip is useful for colony picking (*Figure 3*).

(b) Sample a methylcellulose plug of ~5 μl for each set and prepare a saline control.

(c) Pellet cells 5 min. in microcentrifuge on full speed. The entire 250 μl supernatant is removed. Analysis by PCR mandates extremely careful technique. Use fresh tips for every manipulation to avoid cross-contamination.

(d) Prepare DNA by quick lysis procedure (*Protocol 22* and *Table 2*) using 30–50 μl lysis buffer/sample.

(e) Transduction efficiencies for clonogenic progenitors derived from LTC-IC range from of 20–80% with co-cultivation procedures, and 55–73% with cell-free transduction over bone marrow stroma.

Protocol 26. *Continued*

3. Alternatively, transplant the transduced cells into scid mice (*Protocol 22*). Analyse tissues or human-derived colonies as above.

4. Use G418 selection if the vector encodes *neo*.

(a) This requires dose-response curves on each selected lot of G418 to eliminate 'break-through' colonies. Some lots are unsuitable because of excessive non-specific toxicity at the required doses.

(b) It may not reflect true 'marking' due to possible down-regulation of expression related to the position of integration.

Acknowledgements

The authors wish to acknowledge Denise Simoneaux and John Cannon for critical help in developing this manuscript, S. Kamel-Reid and J. E. Dick for helpful discussions regarding the immune deficient mice, as well as all individuals in the Belmont and Caskey laboratories, past and present, that have been instrumental in the development of these techniques. This work was supported by grants USPH 5 RO1 A130243–02/05 and R29 HD22880 from the National Institutes of Health and the Howard Hughes Medical Institute.

References

1. Joyner, A., Keller, G., Phillips, R. A., and Berstein, A. (1983). *Nature,* **305,** 556.
2. Williams, D. A., Lemischka, I. R., Nathan, D. G., and Mulligan, R. C. (1984). *Nature,* **310,** 476.
3. Dick, J. E., Magli, M. C., Huszar, D., Phillips, R. A., and Berstein, A. (1985). *Cell,* **42,** 71.
4. Dainiak, N. (1991). *Blood,* **78,** 264.
5. Jordan, C. T. and Lemishchka, I.R. (1990). *Genes Dev.,* **4,** 220.
6. Turhan, A. G., Humphries, R. K., Phillips, G. L., Eaves, A. C., and Eaves, C. J. (1989). *N. Engl. J. Med.,* **320,** 1655.
7. Dexter, T. M. and Spooncer, E. (1987). *Annu. Rev. Cell Biol.,* **3,** 423.
8. Keown, W. A., Campbell, C. R., and Kucherlapati, R. S. (1990). In *Methods in enzymology*, Vol. 185 (ed. D. V. Goeddel), pp. 527–37. Academic Press, San Diego.
9. Keating, A., Horsfall, W., Hawley, R. G., and Toneguzzo, F. (1990). *Exp. Haematol.,* **18,** 99.
10. Kriegler, M. (ed.) (1990). *Gene transfer and expression, a laboratory manual.* Stockton Press, New York, NY.
11. Karlsson, S., Van Doren, K., Schweiger, S. G., Nienhuis, A. W., and Gluzman, Y. (1986). *EMBO J.,* **5,** 2377.
12. LaFace, D., Hermonat, P., Wakeland, E., and Peck, A. (1988). *Virology,* **162,** 483.
13. Williams, D. A. (1990). *Hum. Gene Ther.,* **1,** 229.

14. Markowitz, D., Goff, S., and Bank, A. (1988). *J. Virol.,* **62,** 1120.
15. Markowitz, D., Goff, S., and Bank, A. (1988). *Virology,* **167,** 400.
16. Temin, H. M. (1986). In *Gene Transfer* (ed. R. Kucherlapati), pp. 149–87. Plenum Press, New York.
17. Valerio, D., Einerhand, M. P., Wamsley, P. M., Bakx, T. A., Li, C. L., and Verma, I. M. (1989). *Gene,* **84,** 419.
18. Hantzopoulos, P. A., Sullenger, B. A., Ungers, G., and Gilboa, E. (1989). *Proc. Natl Acad. Sci. U.S.A.,* **86,** 3519.
19. Moore, K. A., Scarpa, M., Kooyer, S., Utter, A., Caskey, C. T., and Belmont, J. W. (1991). *Hum. Gene Ther.,* **2,** 307.
20. Scarpa, M., Cournoyer, D., Muzny, D. M., Moore, K. A., Belmont, J. W., and Caskey, C. T. (1991). *Virology,* **180,** 849.
21. Bradley, T. R. and Metcalf, D. (1966). *Austr. J. Exp. Biol. Med.,* **44,** 287.
22. Pluznik, D. H. and Sachs, L. (1965). *J. Cell. Physiol.,* **66,** 319.
23. Fauser, A. A. and Messner, H. A. (1978). *Blood,* **52,** 1243.
24. Dexter, T. M. (1989). *Br. Med. Bull.,* **45,** 337.
25. Suda, T., Suda, J., and Ogawa, J. (1983). *Cell. Physiol.,* **117,** 308.
26. Leary, A. G. and Ogawa, M. (1987). *Blood,* **69,** 953.
27. Bradley, T. R., Hodgson, G. S., and Bertoncello, I. (1980). In *Experimental haematology today* (ed. S. J. Baum, G. D. Ledney, and A. Kahn), pp. 285–97. Karger, New York.
28. Metcalf, D. (ed.) (1977). *Haemopoietic colonies.* In vitro *cloning of normal and leukaemic cells.* Springer-Verlag, Heidelberg.
29. Sutherland, H. J., Eaves, A. C., and Eaves, C. J. (1991). In *Bone marrow processing and purging: a practical guide* (ed. A. P. Gee), pp. 155–71 CRC Press Inc, Boca Raton, FL.
30. Jagannath, S., Reading, C. L., Dicke, K. A., Tindle, S., Devaraj, B., Tucker, S. L., and Spitzer, G. (1987). *Bone Marrow Transplant.,* **1,** 281.
31. Dexter, T. M., Allen, T. D., and Lajtha, L. G. (1977). *J. Cell. Physiol.,* **91,** 335.
32. Gartner, S. and Kaplan, H. S. (1980). *Proc. Natl Acad. Sci. U.S.A.,* **77,** 4756.
33. Sutherland, H. J., Eaves, C. J., Eaves, A. C., Dragowska, W., and Landsdorp, P. M. (1989). *Blood,* **74,** 1563.
34. Winton, E. F. and Colenda, K. W. (1987). *Exp. Haematol.,* **15,** 710.
35. Sutherland, H. J., Lansdorp, P. M., Henkelman, D. H., Eaves, A. C., and Eaves, C. J. (1990). *Proc. Natl Acad. Sci. U.S.A.,* **87,** 3584.
36. Fraser, C. C., Eaves, C. J., Szilvassy, S. J., and Humphries, R. K. (1990). *Blood,* **76,** 1071.
37. Eaves, C. J., Cashman, J. D., and Eaves, A. C. (1991). *J. Tiss. Cult. Meth.,* **13,** 55.
38. Schmitt, R. M., Bruyns, E., and Snodgrass, H. R. (1991). *Genes Dev.,* **5,** 728.
39. Wiles, M. V. and Keller, G. (1991). *Development,* **111,** 259.
40. McMahaon, A. P. and Bradley, A. (1990). *Cell,* **62,** 1073.
41. Till, J. E. and McCulloch, E. A. (1961). *Radiat. Res.,* **14,** 213.
42. Ploemacher, R. E. and Brons, N. H. C. (1989). *Exp. Haematol.,* **17,** 263.
43. Fletcher, F. A., Williams, D. E., Maliszewski, C., Anderson, D., Rives, M., and Belmont, J. W. (1990). *Blood,* **76,** 1098.
44. Bodine, D. M., Karlsson, S., and Nienhuis, A. W. (1989). *Proc. Natl Acad. Sci. U.S.A.,* **86,** 8897.

45. Fletcher, F. A., Moore, K. A., Ashkenazi, M., De Vries, P., Overbeek, P. A., Williams, D. E., and Belmont, J. W. (1991). *J. Exp. Med.,* **174,** 837.
46. Moore, M. A. S. (1991). *Blood,* **78,** 1.
47. Visser, J. W. M. and Van Bekkum, D. W. (1990). *Exp. Haematol.,* **18,** 248.
48. Vries, P. de, Pronk, G. J., and Visser, J. W. M. (1988). *Blood Cells,* **14,** 561.
49. Spangrude, G. J., Heimfield, S., and Weissman, I. L. (1988). *Science,* **241,** 58.
50. Ploemacher, R. E. and Brons, N. H. C. (1988). *J. Cell. Physiol.,* **136,** 531.
51. Bosma, M. J. and Carroll, A. M. (1991). *Annu. Rev. Immunol.,* **9,** 323.
52. Dick, J. E., Lapidot, T., and Pflumio, F. (1991). *Immunol. Rev.,* **124**, 25.
53. Lapidot, T., Pflumio, F., Doedens, M., Murdoch, B., Williams, D. E., and Dick, J. E. (1992). *Science,* **255,** 1041.
54. Karlsson, S. (1991). *Blood,* **78,** 2481.
55. Moore, K. A., Deisseroth, A. B., Reading, C. L., Williams, D. E., and Belmont, J. W. (1992). *Blood,* **79,** 1393.
56. Cournoyer, D., Scarpa, M., Mitani, K., Moore, K. A., Markowitz, D., Bank, A., Belmont, J. W., and Caskey, C.T. (1991). *Hum. Gene Ther.,* **1,** 203.

4

Production of chimeras and genetically defined offspring from targeted ES cells

VIRGINIA PAPAIOANNOU and RANDALL JOHNSON

1. Introduction

Mammalian embryos are extremely resilient in the early stages of their development and can not only tolerate abuse or loss of tissue, but can also functionally incorporate cells from other embryos. This property of mammalian embryos, the ability to incorporate foreign cells during development, has been exploited for several purposes including the elucidation of cell lineages, the investigation of cell potential, and the perpetuation of mutations produced in embryonic stem cells by gene targeting. The extent of contribution of the foreign cells will depend on their normality, their genotype, their mitotic and developmental potential, and their developmental synchrony with the host embryo. If transmission of an altered genotype to the next generation is the goal, the foreign cells must also have the capacity to undergo meiosis and gametogenesis.

Cells from two different mammalian embryos were first combined experimentally to produce a composite animal, dubbed a chimera, more than three decades ago. Pairs of pre-implantation cleaving embryos were mechanically associated *in vitro* until they aggregated together to make single large morulae; these in turn resulted in chimeric offspring (1). Genetic markers were used to distinguish the contributions of the two embryos in these animals. Since then, various methods for making chimeras have been explored to address different types of questions (2). In 1972 it was reported that highly asynchronous embryonic cells, which had been cultured *in vitro*, could contribute to chimeras upon re-introduction into pre-implantation embryos (3). Not long afterward, several groups working with teratocarcinomas, tumours derived from germ cells of the gonad, discovered that stem cells from these tumours, known as embryonal carcinoma cells, could contribute to an embryo if introduced into pre-implantation stages (4–6). It appeared that the undifferentiated stem cells of the tumour had enough features in common with early embryonic cells that they could respond to the embryonic environment, differentiating in a normal manner, even after long periods *in vitro*. Their embryonic potential was not unlimited, however, and many teratocarcinoma cell lines produced tumours in chimeras or made only meagre contributions to the developing fetus (7). Either their derivation from tumours or their extended sojourn *in vitro* rendered these cells so dissimilar from early embryonic cells that they rarely, if ever, had full embryonic potential. Although specific genetic mutations could be selected in these cell lines, none was ever propagated to the next generation through a chimera.

All of this changed dramatically in 1981 when cell lines were derived directly from pre-implantation embryos and maintained *in vitro* in an undifferentiated state (8, 9). These primary cell lines, called embryonic stem (ES) cell lines, correspond closely to cells of the inner cell mass of the blastocyst. Indeed, even after extended periods of culture, ES cell lines can remain multipotential and participate in the formation of all tissues of chimeras, including the gametes. The past five years have seen the rapid exploitation of ES cells to propagate mutations created by gene targeting *in vitro*, taking advantage of the possibility of transmitting the altered gene through the germline of a chimera (10).

In this chapter we will detail standard methods, suitable for use with ES cells, that are used for producing chimeric mice. For the most part, these methods are identical to procedures developed for and used extensively in experimental embryology (11, 12) and their application to ES cell work has been described earlier in this series by Bradley (13). Commonly used markers of chimerism will be discussed and test-breeding procedures will be outlined. We will also describe simple breeding schemes to propagate mutant alleles in generations subsequent to the founder generation and discuss special problems associated with mutants of unknown or unpredictable phenotype. For additional information on mouse husbandry and handling see Hetherington (14). As the use of gene targeting techniques has proliferated, not all laboratories have met with uncomplicated success. We will discuss and evaluate some of the factors that theoretically and practically might enhance the successful germline transmission of the ES cell genotype from chimeras.

2. The starting material

2.1 ES cells

The derivation of ES cell lines directly from embryos was an extension of a large body of work on the stem cells of teratocarcinomas. These rare germ cell tumours are most prevalent in the testes of 129 mice and in the ovaries of LT mice. In determining the genetic basis for this propensity to form germ cell tumours, Stevens developed substrains of 129 mice (129/Sv and 129/Sv-*ter*) with enhanced rates of testicular teratocarcinoma formation (15). In the

Table 1. Allelic differences in marker genes for some strains and substrains commonly used in ES cell experiments, and the ES cell lines derived from them.

Substrain	Locus				ES cell lines	Reference
	Agouti	Albino	Pink-eyed dilution	GPI-1		
129/J	A^w	c, c^{ch}	p	Gpi-1^a		
129/SvJ	A^w	c, c^{ch}	p	Gpi-1^a	PJ1-5	19
129/Sv	A^w	$+^c$	$+^p$	Gpi-1^a	D3	20
129/Sv//Ev	A^w, A	$+^c$	$+^p$	Gpi-1^a or Gpi-1^c	AB-1, CP-1 CCE,CC1.2	21 [a], 22 22, 23
(129/Sv x 129/SvJ)F_1	A^w	$c/+^c$	$p/+^p$	Gpi-1^a	R1	24
129/Ola	A^w	c^{ch}	p	Gpi-1^a	E14	25
C57BL/6J	a	$+^c$	$+^p$	Gpi-1^b	ES632	20
BALB/c	A	c	$+^p$	Gpi-1^a		

[a] and A. Bradley, personal communication.

attempt to derive embryonic stem cell lines directly from embryos, these strains were the obvious choices and some of the first ES cell lines were produced from 129 embryos (8). Although the methods have been extended to embryos of other inbred and random-bred stains (9, 16, 17), 129 ES cell lines remain the most widely available and commonly used. As shown in *Table 1*, however, several 129 substrains, differing for known genetic markers, have been used in deriving these lines.

2.1.1 The parent cell line

Dutiful tissue culture is critical to maintaining an ES cell line and preserving its capacity to contribute to the germline. Cells should be kept at high density, passaged every third day, as a rule, in order to minimize overt differentiation. It is important to avoid passaging cells in clumps, since these clumps will differentiate readily, especially on their outer borders. This differentiation, usually into endoderm, can often be seen as a ring of rounded, diffractile cells around the edges of large ES cell colonies. ES cell lines or clones which have a large percentage of differentiated cells generally contribute poorly to chimeras following embryo injection. However, we have found that clones which have already differentiated extensively can be rescued by selecting colonies of undifferentiated cells from the differentiated culture. This is done much as though one were picking clones in a selection experiment, but the undifferentiated cells are pooled for subsequent culture.

ES cell culture should also include monitoring for the presence of mycoplasma, which can be an insidious spoiler of chimera formation and germline transmission. Mycoplasma are common contaminants of tissue culture cells, and are often found in every culture grown in a laboratory, transmitted from culture to culture by contaminated tissue culture facilities and/or poor sterile technique. Even the most fastidious culturist may pick up mycoplasma from contaminated reagents (such as fetal calf serum or trypsin), so a monthly check for the presence of mycoplasma is good practice. There are several ways to detect the various species of mycobacterium. For routine monitoring, we use the Gen-Probe kit, which utilizes a labelled probe for the presence of mycobacterium-specific ribosomal RNA in tissue culture medium. As an alternative or for confirmation of the presence of mycoplasma contamination, we also use a Hoechst stain on fixed cells, which allows visualization of mycoplasma under a fluorescence microscope (see 18 for detailed protocol); they appear as a thousand points of light around the stained nucleus. Once a mycoplasma contamination is detected, it may be tempting to rescue the contaminated cells using antibiotics, such as gentamicin, effective against mycoplasma. In general, however, while it is possible to cure an ES cell line, the uncertainty about the effect of the mycoplasma or treatment on the cells' pluripotency makes the best solution an immediate disposal of all contaminated cultures and frozen cells. This rule should hold true for all contaminations with bacteria or yeast, as it is usually safest to simply start over once a culture has been invaded. It is for this reason that we routinely culture without antibiotics, since they can mask a contamination by keeping it at a low level. Antibiotics can be added to ES cell cultures which are deemed irreplaceable, such as initial expansions of targeted clones; good sterile technique should make them unnecessary for routine culture.

Clones from all of the ES cell lines listed in *Table 1* have been used to make germline chimeras. It is nevertheless important, when starting from a frozen aliquot obtained from another laboratory or when working with an untested ES cell line, to ensure that the parent cell line to be used for experiments can contribute to the germline. Chimera production with the parent line is thus the first experiment that should be undertaken by a laboratory beginning to do gene targeting. Positive results will indicate a good cell line as well as adequate cell culture and embryo manipulation techniques.

2.1.2 Feeder cells and media supplements

As discussed in Chapter 2, the routine culture of ES cells requires either monolayers of inactivated feeder cells and/or media supplements to prevent differentiation. The feeder layers can be one of two types: murine embryonic fibroblasts (EMFI) or STO cells (a transformed murine embryonic fibroblast line). Media supplements used include Buffalo rat liver cell (BRL) conditioned medium or leukaemia inhibitory factor (LIF), also called HILDA or DIA, sold commercially as a purified, bacterially expressed recombinant protein (Gibco/BRL, ESGRO). In making a choice of what to use, one has to be aware of the difficulties as well as the benefits associated with each. We have found supplements alone are unsatisfactory for retaining pluripotency of ES cells. The recombinant form of LIF is expensive, doubling the cost of each bottle of ES cell medium. BRL conditioned medium is less expensive, but no less labour intensive than culturing with feeder layers, since stocks of BRL cells must be grown up to make conditioned medium, which is then used to supplement ES medium at a 1:2 or 1:3 dilution.

Feeder cell layers are another possibility and each type has its various adherents. They can be used alone or in combination with media supplements, as we commonly do in our laboratory. The STO cell line is easily available and easy to grow. Since it is a transformed line, it can in principle be expanded indefinitely. STO lines can also be transfected and selected for the expression of exogenous genes, such as LIF or neomycin resistance, which can make growing and selecting ES cell clones easier. Primary embryonic fibroblasts are another commonly used type of feeder cell. Their main disadvantage is that they will not grow indefinitely and thus must be rederived periodically. However, they are very effective in maintaining ES cell pluripotency. Neomycin resistant primary embryonic fibroblasts can be obtained from certain transgenic mouse strains. For further information on feeders and supplements, see Chapter 2.

2.1.3 Drug selection and cloning

Selection protocols vary with the selectable marker employed (see Chapter 1). Germline transmission from selected clones is certainly possible but this is not to say that ES cells are completely unaffected by selection agents. Care should be taken to ascertain that the selection protocol uses the minimum amount of drug necessary to select transfectants. Once they are subcloned, cells should not be maintained under selection conditions. A careful initial isolation will prevent non-clonal isolation of non-resistant cells and allow the selected cells to be cultured without drugs. After clones are characterized, they should only be expanded enough to allow freezing of a reference stock in addition to a seed stock for blastocyst injections. This minimal expansion will help limit the total number of passages of the clone. The reference stock should include one or two aliquots of 10^6 cells, and aliquots for injection should have enough cells to thaw directly into a single well of a 24 well plate (see Chapter 2 for details). When confluent, there will be enough cells for a day's injections, with approximately one quarter of the cells passaged into another well for later injections.

2.2 Mice: setting up for embryo recovery and transfer

C57BL/6J mice are widely used and commercially available and differ from 129 mice in coat colour and at other genetic loci which are useful as markers (*Table 1*). Their embryos proved early on to be compatible hosts for 129 ES cells; chimera formation and most importantly germline transmission could be efficiently obtained. They are by no means the only possible choice of host embryo for ES cell chimeras, and considerations of availability, background genotype, and ease of obtaining large numbers of embryos may also come into play. Although random-bred mice can be used, genetically defined host embryos with several genetic markers that can be used to detect chimerism are usually preferable. In addition, CD-1 random-bred mice generally gave poor chimeras in our hands with minimal contribution to the germline.

In choosing a strain for embryo transfer recipients, the primary consideration should be its reproductive performance, including maternal behaviour. For this reason, commercially available random-bred mice, that have been selected for fecundity, make good embryo transfer hosts (e.g. CD-1, MF-1). Likewise, a random-bred strain is suitable for the vasectomized males used to produce pseudopregnancy in the embryo recipients. Any vigorous strain will do but it is useful to incorporate a coat colour marker that would distinguish offspring of the vasectomized male from the experimental offspring, in the unlikely event of a reanastomosis of the vas deferens following vasectomy.

2.2.1 Selection for oestrus

The oestrous cycle in mice is three to four days long with ovulation occurring at approximately the midpoint of the dark period of a light/dark cycle, although this can be affected by crowding, male pheromones, and by exogenous hormone treatment. The timing of mating can be controlled by altering the light/dark cycle of the mouse room to provide embryos at a given stage at a convenient time. For most strains, a year-round 14 hour light/10 hour dark cycle, with midnight the middle of the dark period, will provide blastocysts suitable for injection in the late morning of the fourth day. The most straightforward way of obtaining sufficient embryos of known age is to determine the stage of the oestrous cycle and mate animals on the night ovulation is predicted. In practical terms, this will require a bank of proven fertile, stud males, set up in individual cages, and a stock of females that can be selected for overnight mating when they are in oestrous or pro-oestrous. External vaginal changes are reliable signs of progression through the oestrous cycle in female mice (*Table 2*) (26). The actual size of the opening is not a good criterion since this varies considerably from mouse to mouse but the

Table 2. Changing appearance of the vaginal epithelium during different stages of the oestrous cycle in mice.

Stage of cycle			
Pro-oestrus	**Oestrus**	**Metoestrus 1 and 2**	**Dioestrus**
Moist	Dry	Dry	Wet
Pink/red	Pink	White	Bluish-red
Folded	Folded	Flaky	Smooth
Swollen	Swollen	Less swollen	Not swollen

following external vaginal epithelial characteristics should be chosen to select females for mating:

- dry but not flaky
- pink rather than bluish-red or white
- swollen so that the tissue bulges out
- wrinkled or corrugated epithelium on both upper and lower vaginal lips

Females caged together without a male may tend to cycle together which simplifies oestrous selection. The morphological criteria for selection are more obvious in some mice than others, for example, vaginal colour changes are more evident in albino than in pigmented animals. In addition, inbred animals generally breed less well than outbred mice and this will be reflected in a larger proportion of non-cycling females in a given population.

The morning following pairing of oestrus females with stud males, the vagina of the females must be checked to ascertain whether mating took place. This will be indicated by the presence of a white or yellow vaginal plug, which is composed of male secretions from the vesicular and coagulating glands. This plug can appear loose in the vagina or may seal it shut. It may be at the surface or out of sight deep within the vagina. A narrow, stainless steel dental spatula (3 mm wide) is a convenient probe for detecting plugs, which feel harder than the surrounding tissue (*Figure 1*). Vaginal plugs usually

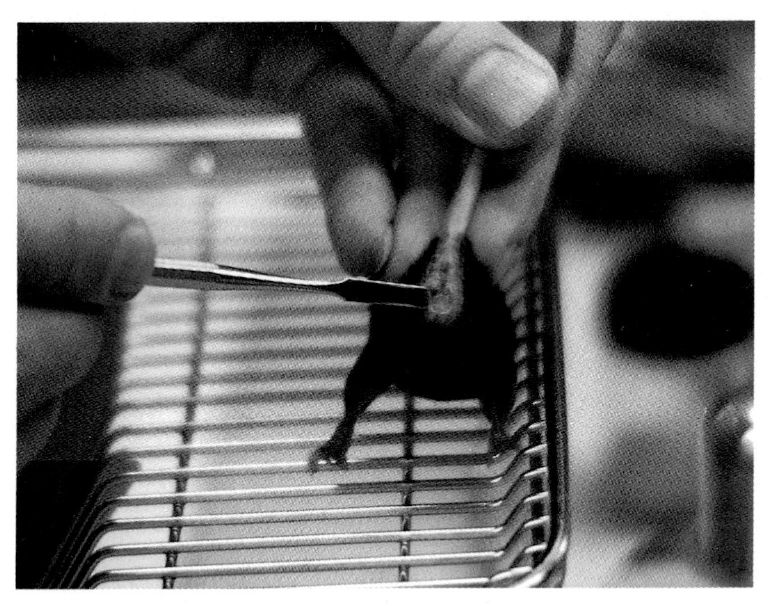

Figure 1. Vaginal plug in a C57BL/6J mouse the morning after mating. A stainless steel dental spatula is used as a probe.

persist for 12 hours after mating, and may even remain for more than 24 hours. Since this is variable, it is wise to check for plugs during the morning after mating. Even with careful selection, one rarely obtains 100% mating, but with experience, the number of animals that will need to be selected for a given number of matings can be determined for each strain being used. In our experience, good stud males are capable of more than one fertile mating per night and of mating three or more times per week, but there are individual and strain variations. A mating record of each stud should be kept to ensure that each male is sexually active. Characteristically, the use of natural matings will entail the maintenance of a relatively large stock of mice.

2.2.2 Superovulation

An alternative to oestrus selection is the hormonal regulation of ovulation with exogenous hormones. In this procedure, a large cohort of immature follicles are induced to mature by an injection of pregnant mare serum gonadotropin (PMSG, Sigma) which has follicle stimulating hormone activity. This is followed 44–48 hours later by an injection of human chorionic gonadotropin (hCG, Sigma) which has luteinizing hormone (LH) activity, thus mimicking the LH surge that normally brings about ovulation. Ovulation will occur approximately 12 hours after hCG administration and should be timed to coincide with the midpoint of the dark cycle. The procedure for superovulation is as follows:

- inject 2–10 IU PMSG intraperitoneally between noon and 16.00 hours on day one
- inject 2–10 IU hCG intraperitoneally at noon on day three
- put females with fertile stud males
- check for vaginal plugs before noon on day four (this is the first day of pregnancy)

Although superovulation would seem the perfect answer to timing experiments and increasing the number of embryos obtained per female used, the theory is far from practical reality. Many variables contribute to inconsistency in the response of the females. Strain and age are important factors, and in some strains immature females (three to five weeks) respond best. Although the number of embryos ovulated should theoretically be related to the dose of hormone, this too is strain and age dependent. Usually only a proportion of the hormonally treated animals mate, although more of them may ovulate. Economy would suggest that the non-mated animals be re-used, but this has only limited success. A recovery period of one to two weeks is necessary for recruitment of another wave of follicles and many females appear refractory to a second dose of hormones, possibly due to the production of antibodies to the foreign hormones.

In spite of these possible frustrations, superovulation is a method well worth a few trials, especially in circumstances where only small numbers of animals are available or large numbers of embryos are needed on a specific day. Even if the number of embryos recovered per animal used is not higher than natural mating, superovulation can synchronize different groups of animals and can, on occasion, result in a bounty of 60–80 embryos from a single mated female.

2.2.3 Embryo transfer hosts

Successful embryo transfer depends on the quality of the embryos and also the suitability of the host maternal environment. Mice are spontaneous ovulators and can be rendered pseudopregnant by mating with sterile males during oestrus, that is, they display the hormonal profile of a pregnant female when the stimulus of mating occurs during oestrus. Embryo transfer at the appropriate time then provides the embryonic signal that leads to the maintenance of the corpora lutea, preventing a return to cyclicity.

Since the window of uterine receptivity to embryo implantation is narrow, and since the uterine environment is hostile to embryos in the first day of pregnancy or pseudopregnancy (when the embryos are normally in the oviduct), the timing and placement of embryos is critical. In general, embryos can be transferred orthotopically to a synchronous host, but since *in vitro* culture and manipulation have the effect of delaying embryonic development, the more efficient method is to transfer embryos to hosts that are one day asynchronous. Thus, fourth day blastocysts are transferred to the uterus of females in the third day of pseudopregnancy (the day of the plug is the first day). Even greater asynchrony can be tolerated, provided it is the embryos that are more advanced and that they are placed in the oviducts. Blastocysts will delay their development during transport to the uterus, and they will still be capable of implantation when the uterus becomes receptive. Thus, embryos at any stage between the one cell and blastocyst can be transferred to the oviducts of host females in the first day of pseudopregnancy and they will implant according to the host's schedule on the fifth day. With this flexibility, host females can be selected for oestrus and set up with sterile males a day or more after the mice providing embryos.

2.2.4 Vasectomy

The final animal component is a stock of sterile males to mate with the embryo transfer hosts to produce pseudopregnancy. For vasectomy and other mouse surgeries detailed in *Protocols 7* and *8*, a thorough knowledge of the anatomy of the reproductive tracts is essential (see references 27, 28). Tribromoethanol anaesthesia (*Protocol 2*) is recommended as a fast acting, safe anaesthetic for all mouse surgeries (30). The following high-quality surgical instruments are necessary for surgery:

- small, sharp scissors
- fine, blunt, curved forceps

- two or three pairs watchmaker's forceps (No. 5)
- iridectomy scissors
- wound clips and applicator (9 mm)

Surgery can be done on an open bench without sterile technique, but antiseptic technique should be observed to minimize the possibility of infection. Instruments should be soaked in 95% alcohol and flamed before use.

Protocol 1. Vasectomy (*Figure 2*)

1. Gather the following equipment:
   - tribromoethanol anaesthesia (see *Protocol 2*)
   - 70% alcohol and cotton
   - surgical silk suture (5.0) and needle
   - surgical instruments
2. Weigh and anaesthetize a mouse and swab the abdomen with alcohol.
3. Make a 1 cm, ventral, midline, transverse skin incision 1 cm rostral to the penis by lifting the skin free of the peritoneum (*Figure 2a*).
4. Make a similar incision through the peritoneum.
5. With the blunt forceps, reach laterally and caudally into the peritoneum and grasp the testicular fat pad, which lies ventral to the intestine and is attached to the testis. Pull it out through the incision, bringing the testis with it.
6. Locate the vas deferens, being careful to distinguish it from the corpus epididymis, and separate it from the mesentery by inserting the closed tips of the iridectomy scissors through the mesentery. Open the scissors and leave them in this position, supporting and isolating a length of the vas deferens (*Figure 2b*).
7. Insert a doubled length of suture under the vas deferens and cut it, leaving two separate pieces (*Figure 2c*).
8. Tie off the vas deferens in two places 2 mm apart and cut out the intervening length (*Figure 2d, e*).
9. Return the testis to the peritoneum and repeat the procedure on the other side.
10. Suture the peritoneal wall and clip the skin incision with one or two small wound clips.
11. Following surgery, leave vasectomized males for at least ten days to clear the tract of viable sperm, and then test mate to ensure sterility before use in experiments.

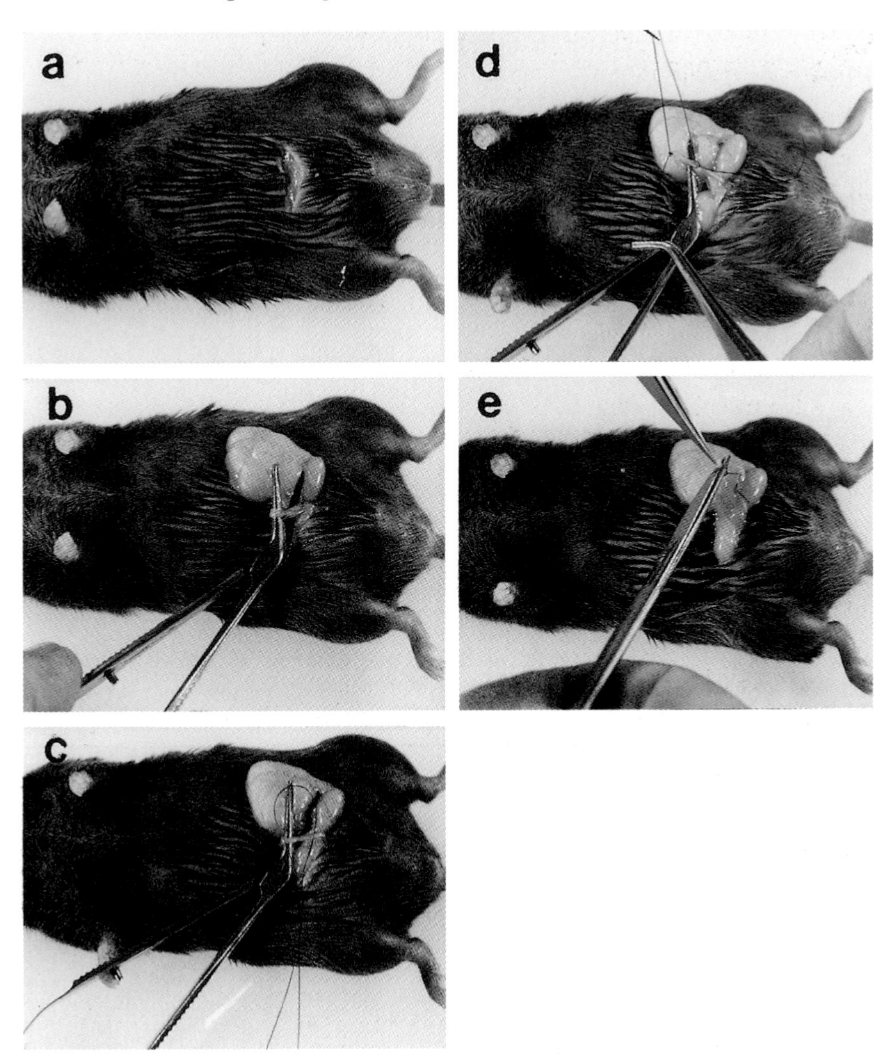

Figure 2. Vasectomy of a male mouse. (a) A transverse incision is made ventrally in the abdominal skin. (b) A similar incision is made in the peritoneum and the testicular fat pad and testis is exteriorized. Iridectomy scissors are used to isolate a short length of the vas deferens. (c) A loop of suture is passed under the vas deferens. (d) After the loop is cut, the two lengths of suture are used to tie off the vas deferens in two places, 2 mm apart. (e) The intervening length of vas deferens is cut out.

Protocol 2. Recipe for tribromoethanol anaesthesia

1. Ingredients:
   - 2.5 g 2,2,2-tribromoethanol (Aldrich)
   - 5 ml 2-methyl-2-butanol (*tertiary* amyl alcohol; Aldrich)
   - 200 ml distilled water
2. Add tribromoethanol to butanol and dissolve by heating (approximately 50 °C) and stirring.
3. Add distilled water and continue to stir until butanol is totally dispersed.
4. Aliquot into brown bottles or foil covered tubes (10 ml) and store in the dark at 4 °C.[a]
5. Warm to 37 °C and shake well before use. The dose for mice is 0.2 ml per 10 g body weight, although up to 1.5 × this dose can be given safely, particularly to fat mice.

[a] Stored properly, this solution is stable for several months. Decomposition to dibromoacetic aldehyde and hydrobromic acid can result from improper storage in light or at high temperatures (above 40 °C). If this occurs, the solution is toxic and will cause death within 24 hours of injection. pH can be used as an indicator of decomposition (a drop of Congo red will turn the solution purple at pH <5), but this test will only be valid if the original solution was pH >5, which can vary considerably with the water source used (29).

2.3 Recovering embryos

2.3.1 Culture media and conditions

Making chimeras necessitates a period of culture *ex vivo*, during which embryos will be in less than optimal conditions. Several culture media have been developed that will maintain the embryos in a viable state and a common one is given in *Protocol 3*. During the time the embryos are kept outside an incubator, a phosphate or Hepes buffer that will maintain pH in room atmosphere is used. If the embryos are to be incubated for any period of time, they should be placed in bicarbonate-buffered medium and kept in 5% CO_2 in air. It should be noted that optimal culture conditions for pre-implantation embryos and for ES cells are not the same so that in making chimeras there is a compromise while the cells are being introduced into the embryo. We routinely use PB-1 + 10% serum (*Protocol 3*) for recovery and handling of embryos and DMEM/Hepes (Sigma) + 20% serum for ES cell injection, although the embryos can equally well be recovered and injected in the latter. There is no evidence to indicate how these various conditions might ultimately affect the contribution of ES cells to chimeras. Keeping embryos cool (10 °C) during manipulations by means of a cooling stage has no adverse effects on their subsequent viability and may facilitate handling of the ES cells during injection.

Protocol 3. Modified PB-1 medium with 10% fetal calf serum for embryo recovery and handling

1. Make up the following stock solutions and mix the indicated volumes:

	Stock solutions (g/100 ml)	Volume (ml)
• NaCl	0.9	68.96
• KCl	1.148	1.84
• $Na_2HPO_4.12H_2O$	5.5101	5.44
• KH_2PO_4	2.096	0.96
• $CaCl_2.2H_2O$	1.1617	0.88
• $MgCl_2$	3.131	0.32
• Na pyruvate	0.020 (in stock NaCl)	22.40

2. Add the following to make up 104 ml of medium:

• penicillin	6.2 mg
• glucose	104 mg
• distilled water	2.16 ml
• phenol red (1%)	0.1 ml

3. To this medium add heat-inactivated fetal calf serum (56°C for 30 min.) at a final concentration of 10%, (e.g. 10 ml/90 ml medium).
4. Filter sterilize, aliquot into sterile containers, and refrigerate until use.

2.3.2 Blastocyst recovery (*Protocol 4*)

Around midday on the third day of pregnancy, embryos traverse the utero-tubal junction. They are at the 8–16 cell stage and will be forming compact morulae. Recovery of morulae from the reproductive tract is covered in Chapter 5. By the middle of the fourth day blastocysts will be present in the uterine lumen and can be recovered by removal and flushing of the uterine horns with the culture media just described.

Protocol 4. Recovery of blastocysts (*Figure 3*)

1. Kill the pregnant animal on the fourth day of pregnancy (three and a half days post-coitus) by cervical dislocation.
2. Swab the abdomen with 70% alcohol and make a small, transverse incision mid-ventrally. Wipe away the cut hairs with alcohol soaked cotton.
3. Gripping the skin anterior and posterior to the incision, tear the skin, pulling it back to the front and hind limbs respectively (*Figure 3a*). Tearing obviates the problem of cut hairs in the incision.
4. Open the peritoneum with a large transverse incision to expose the abdominal cavity.

Protocol 4. *Continued*

5. Locate the reproductive tract dorso-caudally by pushing the intestines cranially. Grasp the point of bifurcation of the uterus, which is near the base of the bladder, with blunt forceps (*Figure 3b*).

6. Lifting the uterus, cut across the cervix and continue to lift, clipping away the mesometrium, the mesentery supporting the uterus, on both sides.

7. Make a cut through the bursa between the oviduct and the ovary or through the utero-tubal junction on each side, freeing the entire tract (*Figure 3c*).

8. Now trim the tract prior to flushing using a low power dissecting microscope.

(a) Spread the tract out in a dry sterile Petri dish and trim away any remaining mesentery and fat.

(b) Cut off the oviduct at the utero-tubal junction (if this was not already done) and make a 0.5 cm longitudinal clip in the end of each horn.

(c) Cut across the cervix to expose the entrances to both uterine horns.

9. Rinse with a few drops of medium and place the tract into a sterile watchglass. Using a rubber bulb on a Pasteur pipette which has been pulled out and broken off, flush approximately 0.5 ml of medium (*Protocol 3*) through each horn by inserting the pipette tip into the cervical end. The uterus will balloon slightly and medium should flow freely through the tract.

Embryos will settle to the bottom of the flush dish very quickly. It is useful to have round-bottom watchglasses (Carolina Biological) to flush the uteri so that the embryos can be swirled to the centre of the dish. The cleaner the dissection of the uterus, the less cellular debris, fat, and blood will obscure the visualization of the blastocysts (*Figure 4*). Embryos can be collected from the flush dish and transferred to a watchglass containing fresh medium with a drawn-out, mouth-controlled Pasteur pipette prior to their introduction into the micromanipulation chamber or dish.

3. Introduction of ES cells into embryos

3.1 Micromanipulation apparatus

3.1.1 Microscopes

ES cell injections can be done under phase, bright field, or differential interference contrast optics. The microscope, which should be located on an anti-vibration table, should have at least one low and one higher power objective giving a magnification range of approximately 63 × to 200 ×. Another very useful feature is a fixed stage so that the embryos and micro-

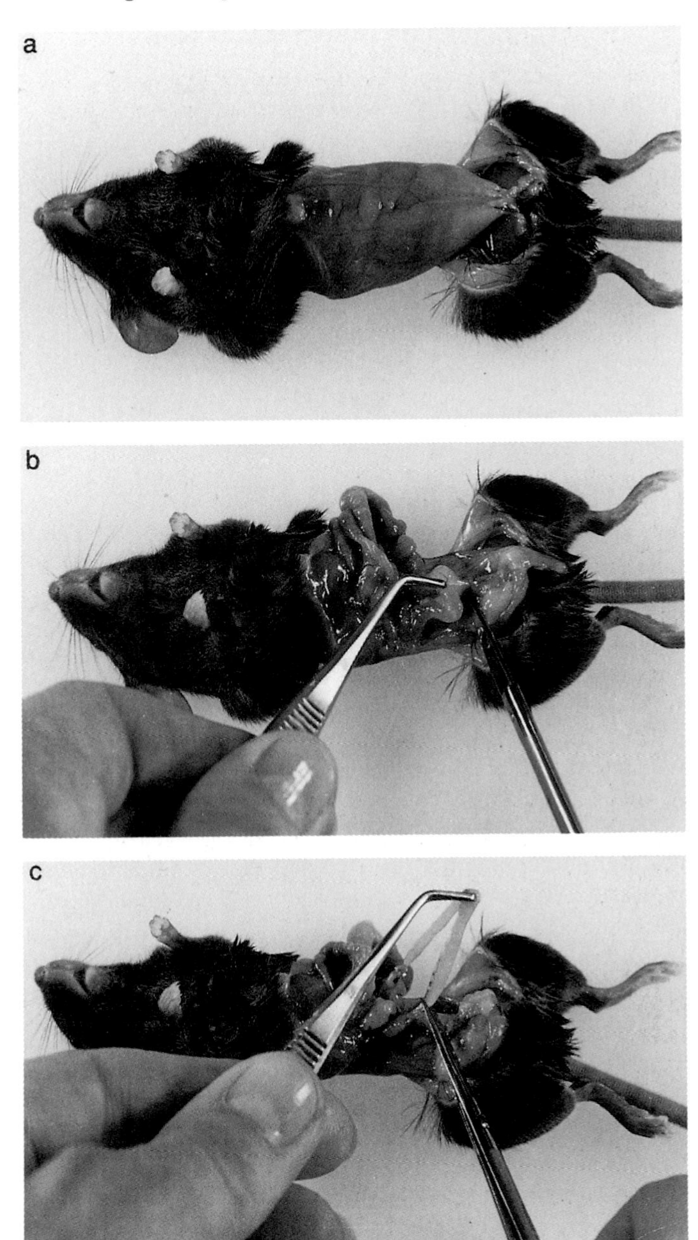

Figure 3. Blastocyst recovery at three and a half days post-coitus. (a) The peritoneum is exposed by tearing the skin. (b) After the peritoneum is opened and the intestines are pushed cranially, the bifurcation of the uterine horns is grasped and a cut is made across the cervix. (c) The uterus is lifted and each horn is freed by cutting the mesentery and finally by cutting at the utero-tubal junction, as shown here, or between the oviduct and ovary if the oviducts are to be recovered.

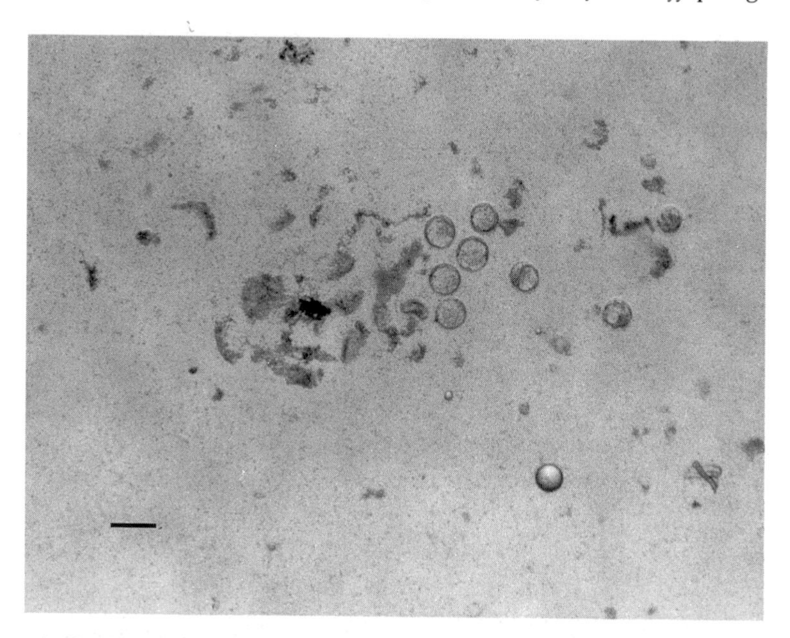

Figure 4. Blastocysts immediately after being flushed from the uterus. Blood, cellular debris, fat, and oil drops can make embryos difficult to see. Bar = 0.1 mm.

instruments remain in the same focal plane relative to each other when the microscope focus is adjusted. The convenience of this feature is well worth the effort of modifying a microscope if a fixed stage is not available. However, if stage-mounted micromanipulators are used, this is not a concern.

Another useful addition to the microscope is a cooling stage. Its main advantage is to prevent stickiness during cell injections and to impart a degree of rigidity to cells and embryos. Any cooling device that will allow free access to the stage and will also keep the injection chamber at 10°C will suffice (13). We use a Physitemp cooling stage with a modified stage to accommodate either an inverted or standard microscope (*Figure 5*).

There are several microscope configurations that can be used for blastocyst

---

Figure 5. Micromanipulation set-up with Leitz micromanipulators and (a) an inverted microscope or (b) a standard microscope. Both are placed on pneumatic, anti-vibration tables. (a) For the inverted microscope, the manipulators are raised to the level of the stage on metal supports. A Gilmont micrometre syringe can be seen mounted on the *right*. Instrument holders are positioned for use in a culture dish. Note that the manipulators are level and the instrument holders are tilted at the ball and socket joint to attain the correct angle. (b) In the set-up with the standard microscope, the manipulators are mounted on a Leitz baseplate. On the *right* is a Narishigi micrometre syringe used for the holding pipette, and on the *left* is the Stoelting micrometre for cell injection. The control box for the Physitemp cooling stage is on the *left*.

injections. An inverted microscope (*Figure 5a*) requires that the injections be done on the bottom of a culture dish (or lid) or a depression slide. The embryos and cells can either be placed in a large volume of culture medium, or they can be placed in microdrops of medium covered with a layer of inert oil such as heavy liquid paraffin oil (HLP, Fisher). Microdrops will allow for

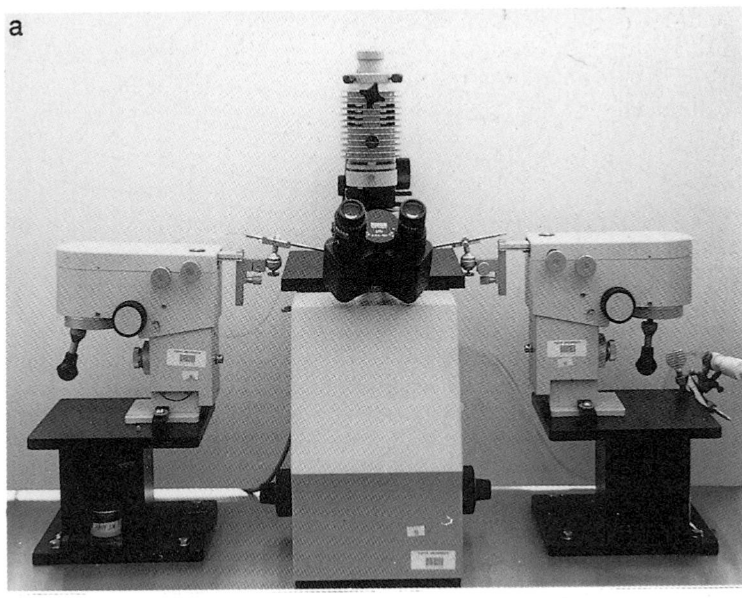

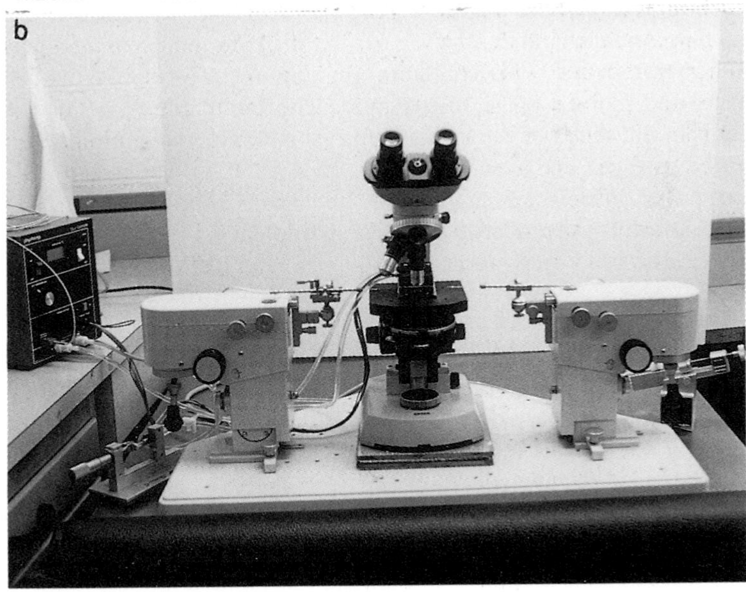

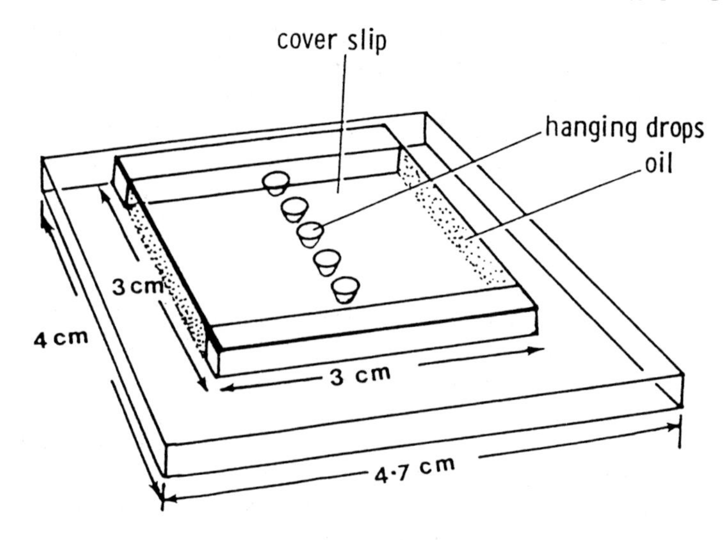

Figure 6. Diagram of a hanging drop micromanipulation chamber (Leitz). The chamber is made of glass with two parallel supports on which a coverslip will rest. To load the chamber, petroleum jelly is placed on the top of the supports. Drops of medium are placed in a row on an acid-alcohol cleaned, sterilized glass coverslip which is then inverted on to the supports. Finally, the space under the coverslip is filled with HLP. A chamber can be made by glueing the cut off ends of a Pasteur pipette on to a glass slide (reproduced from 12).

separation of different groups of embryos or cells and also obviates the problem of searching for the embryos in a large field. An alternative microscopic configuration is the standard microscope (*Figure 5b*) and the use of a micromanipulation chamber (*Figure 6*) in which the embryos and cells are kept in hanging drops. This configuration requires a long-working-distance condenser and a fixed stage microscope. The main difference in the two methods is in the shape of the instruments used for cell injection, and the fact that in one case the embryos are resting on a solid surface and in the other they are resting on an interface of oil and medium. Other than this, the choice may be dictated by the microscope and manipulators available, since both methods can be used quite successfully.

3.1.2 Micromanipulators

Several types of micromanipulator are available that are suitable for ES cell injection. Cell injection into blastocysts is a fairly large-motion procedure so a wide range of movement, particularly in the coarse adjustments, is important. For this reason, stage-mounted manipulators are not as convenient as those that are free-standing. The precision and working range of movement needed for the holding pipette is less than for the cell injection pipette so that a simple manipulator could be used for that instrument. However, we recommend two Leitz micromanipulators (one right and one left handed model; Leitz Instruments) which satisfy all the requirements for ES cell injection and can be used for virtually any other micromanipulation as well. The Leitz manipulator is a direct mechanical device with an XY joystick movement with a wide range of adjustment of the movement ratio. Movement in the third plane is by a screw device. All three planes of movement have coarse and fine adjustments and there is flexibility also in the positioning of the instrument holder in all directions.

3.1.3 Micrometre syringes

Mechanical suction-and-force syringe devices are used for both the holding and cell injection pipettes in conjunction with Leitz instrument holders. There is a wealth of choice of micrometre-driven syringe devices, but it is essential that they both push and pull the syringe plunger to create suction and force. It is a mistake to use too fine a device (one with very small displacement per turn), especially for the holding pipette. Both devices are attached to instrument holders with polythene tubing filled with HLP or silicon oil. A three-way valve is placed in the line between the syringe and instrument for easier filling and instrument changing. For the holding pipette, we use either a Gilmont 2 ml micrometre syringe or a Narishige Model IM-58, with a 3 ml syringe, which has the added advantage of a magnetic base. More precision and finer control over suction is needed for the cell injection pipette. For this we either use a deFonbrune Suction-and-Force pump or a Stoelting micrometre syringe assembly (*Figure 7*). This comes with a 50 μl Hamilton syringe that should be changed to a larger size, (e.g. 0.5–1 ml) for better

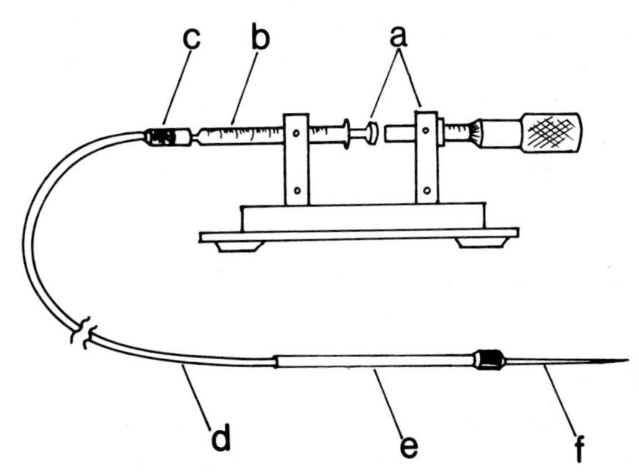

Figure 7. Diagram of a Stoelting micrometre syringe assembly with attached injection pipette. A rubber band connecting the two points indicated by (a) will allow the syringe to be used for suction as well as expelling. The assembly includes a Hamilton microlitre syringe (b) and an adaptor (c). Flexible polythene tubing (d) is fitted to connect the syringe to a Leitz instrument holder (e) into which the injection pipette (f) is inserted. The syringe and tubing are filled with oil. It is useful to have a three-way valve (not shown) in the line (reproduced from 30).

control. A rubber band or spring clip will need to be added to allow retraction as well as pushing of the plunger. The syringes are placed on the opposite side of the manipulator from the instrument they control so that suction or force can be applied at the same time as the instrument is being moved with the joystick.

3.2 Microinstruments

3.2.1 Holding pipettes

Microinstruments for use in blastocyst injection are fashioned out of glass capillary that will fit the instrument holders, (e.g. Drummond custom capillary, 1 mm o.d., 0.75 mm i.d.). Only two instruments are necessary, a pipette to hold and stabilize the blastocyst and a pipette to pick up and inject the cells. The tips of these instruments are fashioned in the same way regardless of the microscope/manipulator configuration used; only the more proximal shaping of the shafts will differ depending on the manipulation set-up.

The overall tip diameter of a holding pipette should be broad enough to stabilize an embryo. The opening should be small enough to prevent gross deformation of the embryo or the zona pellucida (ZP) while still allowing sufficient area for suction to hold the embryo tightly (see *Protocol 5* for the method).

Protocol 5. Making an embryo holding pipette

1. Pull out a length of glass capillary either with a pipette puller (such as Kopf) or by hand over a small flame. We prefer to make this initial pull by hand over a low flame to give a taper of 2–3 cm, which confers better control over the suction than the shorter taper obtained with a pipette puller.
2. Break the pipette cleanly at an overall diameter of 60–90 μm (*Figure 8*) using a microforge (deFonbrune).
   (a) Start with a bead of glass on the filament tip.
   (b) Focus on the pipette and bring the filament into the same focal plane.
   (c) Turn on the filament and touch the molten glass bead to the pipette at the desired diameter just until it sticks, taking care not to melt the pipette.
   (d) Turn off the filament and the contraction should break the pipette.
3. Turn on the filament and bring it close to the tip of the pipette to fire-polish and close it down to a diameter of 12–20 μm. The pipette is now ready to use in a hanging drop.
4. Place additional bends in the shaft of the pipette if depression slides or culture dishes are used for manipulation, so that the final orientation of the tip is horizontal when it enters the microscope field (*Figure 9*). Make these bends with the microforge in the narrow part of the shaft (*Figure 10*) or in the thick part of the shaft over the small flame of a microburner that can be made by connecting a 30 G hypodermic needle to a gas supply.

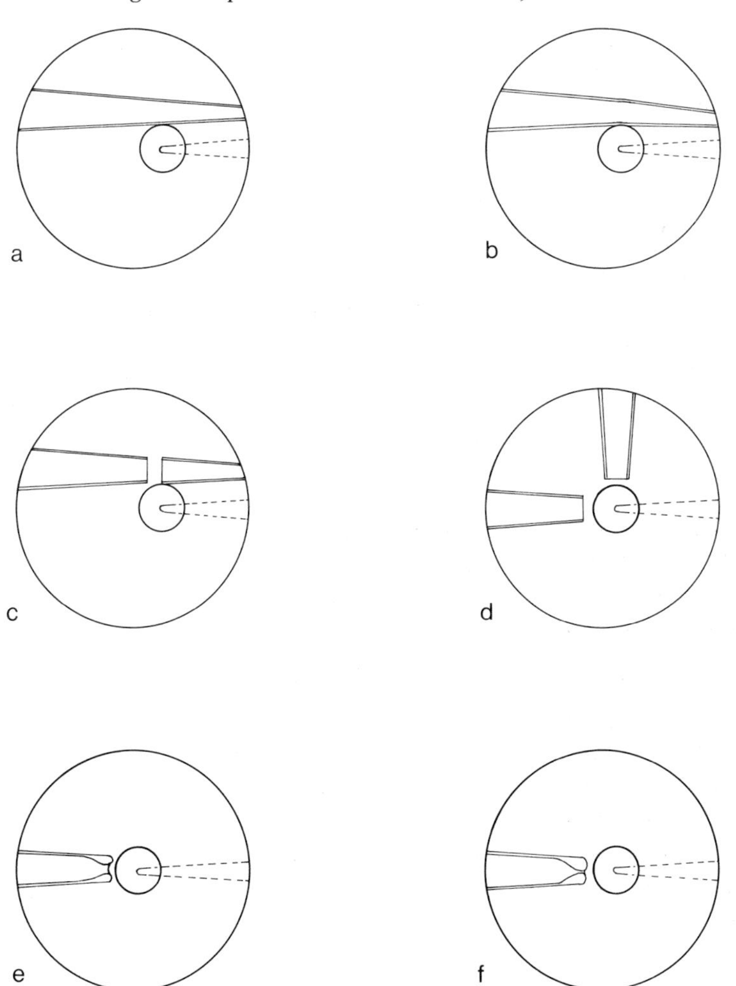

Figure 8. Making an embryo holding pipette. (a) A molten glass bead on the tip of the microforge filament is brought into contact with a pulled-out pipette at the appropriate diameter and (b) the bead fuses with the pipette. (c) The filament will contract when the element is switched off, breaking the pipette. (d–f) The pipette is closed down to a small opening by fire-polishing the tip with a hot filament (reproduced from 13).

3.2.2 Cell injection pipettes

Either blunt or bevelled pipettes can be used for cell injection. Blunt tips require some force to penetrate the trophectoderm layer of the blastocyst while bevelled tips can be inserted more deliberately. Both methods have their proponents and are equally effective for blastocyst injection, although making the bevelled tip requires more steps. Morula injection requires the use of a bevelled pipette to avoid damage to the blastomeres.

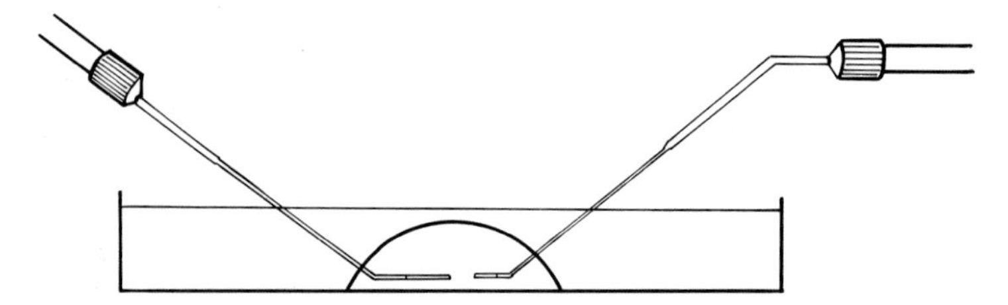

Figure 9. Positioning of instruments for use with an inverted microscope, with bends in the glass pipettes. The pipette on the *right* has two bends which allows the instrument holder to be positioned horizontally, while the one on the *left* has a single bend so that the instrument holder is at an angle. Which position is used will depend on the location of the manipulators with respect to the microscope stage.

For both types of pipette, glass capillary is drawn out to a gradual 1 cm taper with a pipette puller. For a blunt pipette, the tip is broken off using a microforge (see *Protocol 5*, step 2 and *Figure 8a–c*) at an internal diameter of 12–15 μm, a diameter just large enough to allow the entry of single ES cells without deformation. For a bevelled tip, the pipette is placed on a silicon pad under a dissecting microscope and a scalpel blade is used to break off the tip. Pipettes of the correct diameter, with smooth, sharp bevels that are neither jagged nor too long, are then selected for use (13). No fire-polishing of either type of injection pipette is necessary. For use in a culture dish or depression slide, bends must be made in the shaft of cell injection pipettes so the final working position of the tip is horizontal with respect to the microscope stage (*Protocol 5*, step 4; *Figures 9* and *10*).

3.2.3 Embryo transfer pipettes

Making the pipette for transferring embryos into a host female is very similar to making a holding pipette (*Protocol 5*). Capillary is drawn out by hand in the same way and the tip is broken off similarly using a microforge. The difference is that the overall diameter at the tip is larger so that the opening can easily accommodate embryos (110–130 μm internal diameter). For this same reason, fire-polishing of the tip is minimal, just enough to take off the sharp edges without closing down the opening. No bends are necessary (*Figure 11*).

3.3 Injection procedures

3.3.1 Blastocyst injection

On the day of an experiment, blastocysts are recovered from pregnant females in the morning. ES cells in an exponential growth phase are fed; one hour later they are trypsinized (Chapter 2, *Protocol 4*) and maintained as a single cell suspension in PB-1 medium at 4°C. This suspension can be used throughout the day.

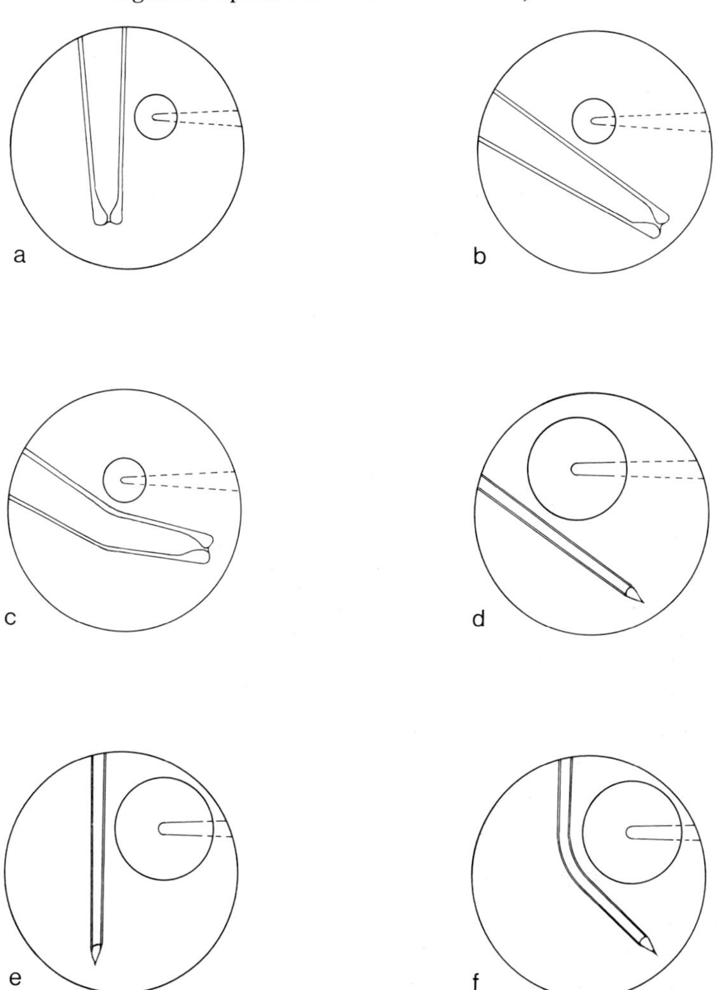

Figure 10. Making bends in a pipette with a microforge. (a–c) Holding pipette. The pipette is oriented as in either (a) or (b) and the filament is heated, causing the pipette to bend towards the filament (c). (d–f) Bevelled injection pipette. If the bend is made near the end of a fine pipette, a higher magnification and lower heat can be used. While making the bend, the bevel should be oriented so that its final position in the injection chamber will be as indicated in *Figure 13* (reproduced from 13).

Blastocysts and a single cell suspension of ES cells are introduced into drops of medium in the manipulation chamber using a drawn-out, mouth-controlled Pasteur pipette. This can be done under a dissecting microscope for a hanging drop chamber, or under the manipulation microscope for a dish or depression slide. When using a hanging drop chamber, set-up the microscope so that the embryos are in focus and then remove the chamber from the

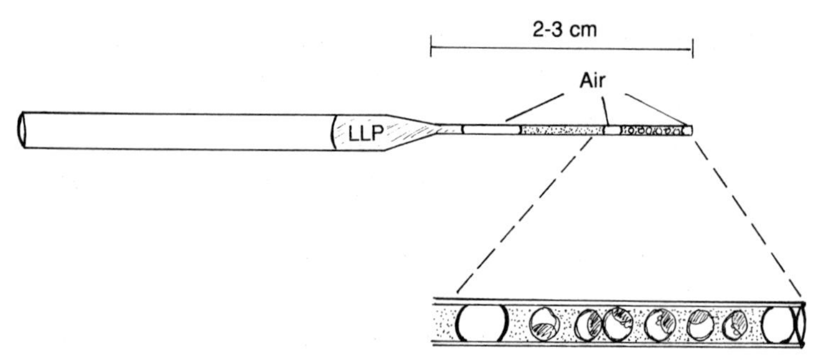

Figure 11. Diagram of embryo transfer pipette ready for embryo transfer. The narrow, pulled-out tip should measure 2–3 cm. In filling the pipette, light liquid paraffin oil is taken up to the thick region, then a column of air to separate oil from medium, then medium, then a small marker air bubble followed by the embryos (six blastocysts are shown here) and, finally, a small bubble at the tip.

microscope. Now the injection instruments can be set-up in the micromanipulators, without fear of breaking them on the chamber, as follows.

(a) Put each instrument into its holder and manipulator, fill with oil by positive pressure.

(b) Without adjusting the microscope focus, roughly position the tips parallel in the field using the coarse adjustments of the manipulator; put them into the same focal plane using the fine adjustments of the manipulators.

(c) Withdraw the instruments far enough to allow the introduction of the manipulation chamber on to the stage and restore the instruments to position.

A dish or depression slide can be left in place while setting-up the micro-injection instruments. The blastocyst injection procedure is outlined in *Protocol 6*.

Protocol 6. Blastocyst injection

A. *Using blunt tipped pipettes* (12, 31; *Figure 12*)

1. Pick up 12–15 small, round ES cells in a tight column in the tip of the injection pipette using suction.
2. Pick up a blastocyst with the holding pipette by suction, positioning it so that the ICM can be seen in profile against the pipette (*Figure 12a*).
3. With the cell injection pipette in the centre of the field, and the blastocyst in focus but offset to the north or south, position the injection pipette so that it is level with the surface of the ICM when it is at the full extent of the range of movement of the manipulator joystick (*Figure 12a*).

4. Retract the injection pipette with the joystick, reposition the blastocyst to the centre of the field, check that the equatorial plane of the blastocyst is in the same focal plane as the injection pipette (*Figure 12b*), then apply force with the joystick to pop the injection pipette through the trophectoderm wall. The pre-set limit on the joystick movement will prevent damage to the ICM (*Figure 12c, d*).
5. Expel the ES cells into the blastocoelic cavity and remove the pipette (*Figure 12e*).
6. Place the injected embryo to one side of the chamber drop and repeat the procedure.

B. *Using a bevelled pipette* (13) (see *Figure 13a*)

1. Follow steps 1 and 2 as above.
2. Put the injection pipette in the same focal plane and slowly penetrate the trophectoderm layer, trying to insert the tip of the pipette at a junction between two cells.
3. Follow steps 5 and 6 above.

The microinstruments will only need changing if they become clogged or sticky. As a rule, we set-up a fresh manipulation chamber with fresh ES cells after several hours or when the pH of the medium changes. When all of the embryos in a chamber have been injected with ES cells, they can be transferred to culture medium for a brief period of culture prior to transfer to pseudopregnant hosts.

3.3.2 Morula injection

Introduction of ES cells into embryos at the morula stage is another means of producing chimeras. The forceful method of blastocyst penetration can not be used for injecting morulae without damage to the blastomeres. Instead, a bevelled pipette is used to penetrate the ZP and cells are inserted just under the ZP (*Figure 13*). Alternatively, the cells can be placed among the blastomeres in an embryo that has not yet compacted, or one that has been decompacted by a brief culture in calcium-free medium. No study has been published that critically compares the success of morula versus blastocyst injection. Using morula injection we have produced ES cell chimeras, however, our experimental results indicate no clear advantage of this method over blastocyst injection.

4. Embryo transfer

Oviduct transfers (*Protocol 7*) and uterine transfers (*Protocol 8*) are similar although oviduct transfers require somewhat greater skill. Since the ostium of the oviduct is enlarged and fairly obvious during the day after ovulation, this is

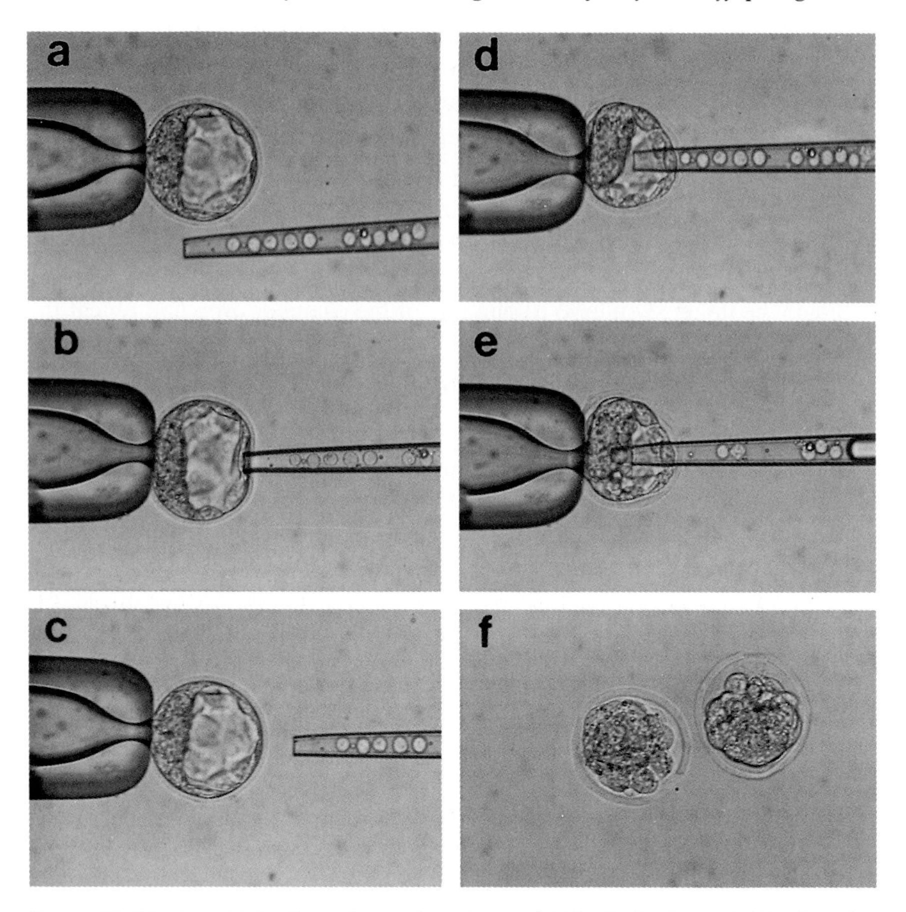

Figure 12. Blastocyst injection using a blunt pipette. (a) The limit of travel of the injection pipette is set when the pipette is in the position indicated. ES cells can be seen in the injection pipette and the blastocyst is held by suction on the holding pipette. (b) The focal planes of the injection pipette and blastocyst can be checked by gently touching the pipette to the embryo. (c) The injection pipette is repositioned with the joystick and (d) forced through the trophectoderm layer by a sudden movement, or bash, of the joystick. (e) Cells are then expelled into the blastocoelic cavity. (f) Injected embryos prior to re-expansion.

a convenient time for oviduct transfers. Uterine transfers should be done on the third or fourth day of pseudopregnancy (see Section 2.2.3 for details on setting up the embryo transfer hosts).

Protocol 7. Embryo transfer to the oviduct

1. Weigh and anaesthetize a pseudopregnant female (see *Protocol 2*).
2. Prepare the embryo transfer pipette by connecting it to a mouth pipette

and filling it with light liquid paraffin oil (Fisher Scientific) at least to the end of the taper. Take up a column of air, medium, a small air bubble, the embryos to be transferred, and finally a small air bubble at the tip (*Figure 11*). Set this aside while the mouse is prepared.

3. Clean the back of the mouse with alcohol and make a 1 cm transverse incision in the skin at the level of the first lumbar vertebra (*Figure 14a*). Wipe the hair away with cotton dampened with alcohol.
4. Slide the skin incision laterally until the left ovarian fat pad and ovary are visible through the peritoneal wall. Make a 3 mm incision through the peritoneum and grasp the fat pad with blunt forceps, pulling it and the ovary and oviduct out through the opening in the direction of the midline.
5. Position the mouse on a dissecting microscope stage with its head facing away from you at 11 o'clock (*Figure 14b*).
6. Stabilize the ovary outside the peritoneum by clamping with a small sepaphim clip or drape the fat pad over a small square of index card with a notch in one side, pulling the ovary, oviduct, and a short length of uterus through the notch (*Figure 14c*).
7. With a pair of fine watchmaker's forceps in each hand and the loaded transfer pipette in the right hand, locate the fimbria of the oviduct by looking in the space between the ovary and the oviduct. Using the forceps, tear a hole in the bursa, the vascular, transparent membrane enclosing the ovary. It may be necessary to remove fluid or blood with a tissue in order to visualize the ostium. Alternatively, a drop of epinephrine can be used to control bleeding (32).
8. Insert the tip of the pipette into the ostium of the oviduct and blow in the embryos along with the air bubbles on either side of them. The orientation of the oviduct opening is such that if the mouse is facing away from you at about 11 o'clock, and the fat pad is pulled toward the midline, the pipette can be inserted from the right, parallel to the backbone for either a right or left oviduct transfer (*Figure 15*).
9. The marker air bubbles will be visible in the first turn of the oviduct but the transfer pipette should also be checked to ensure that all embryos were expelled.
10. Carefully replace the tract into the peritoneum.
11. Repeat steps 4–10 on the right side for a bilateral transfer.
12. Provided the peritoneum incisions are small, they can be left open. Close the skin incision with a small wound clip.

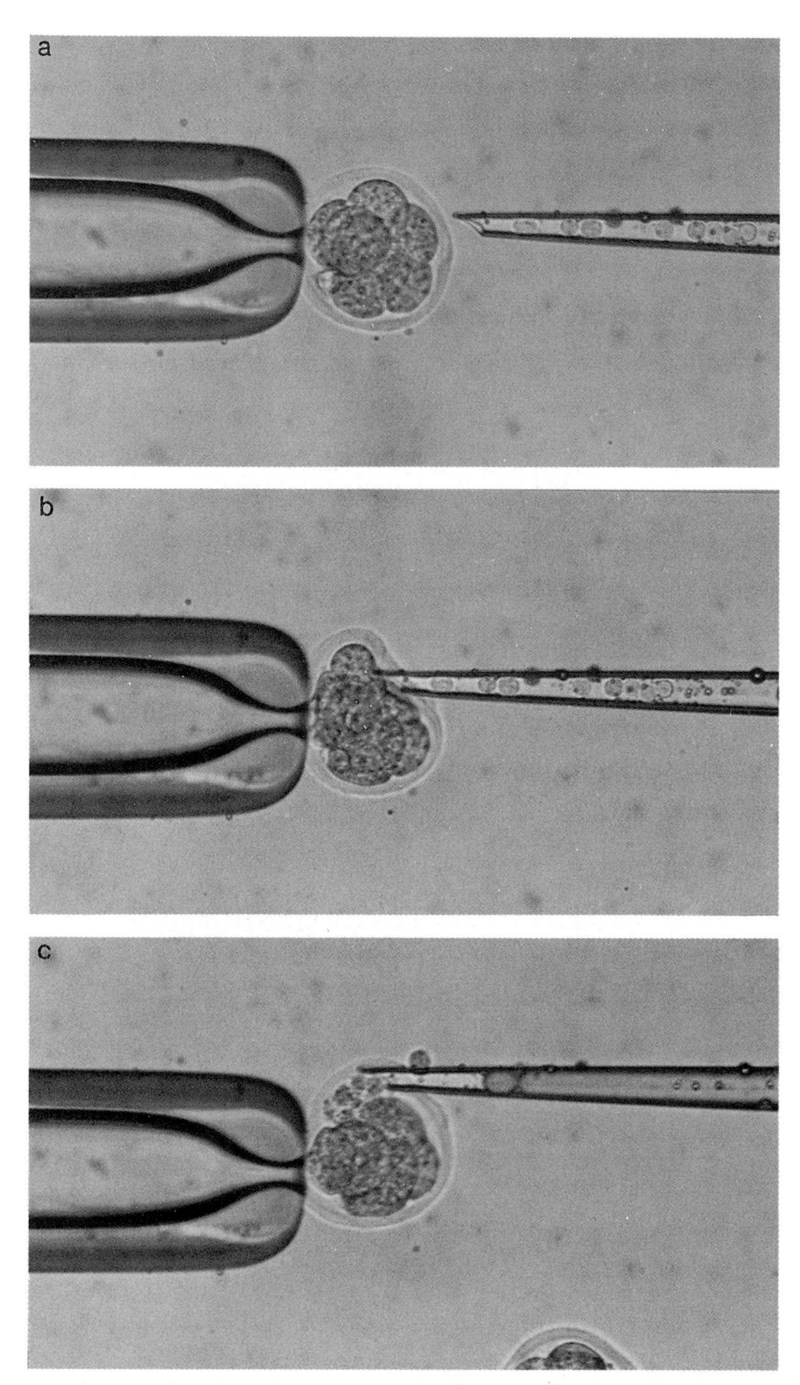

Figure 13. Injection of cells under the ZP of an eight-cell embryo. (a) The embryo is held by suction and ES cells are in the bevelled injection pipette. (b) The tip of the pipette is slowly inserted through the ZP and (c) the cells are expelled into the perivitelline space once the pipette has fully penetrated.

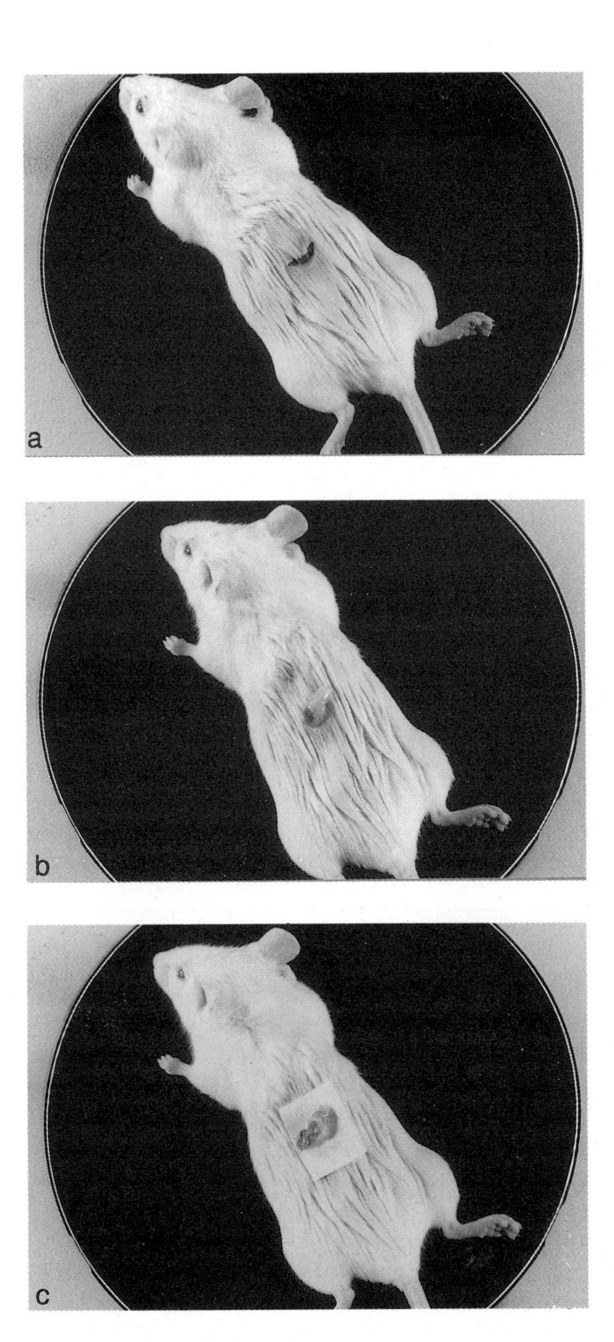

Figure 14. Preparation for embryo transfer. (a) A single dorsal transverse incision will serve for embryo transfer to both the left and the right. The mouse is in the correct position for transfer to either side. (b) The left ovary and ovarian fat have been exteriorized through a lateral peritoneal incision and the skin incision. With the exteriorization of a small, distal segment of the uterus, the mouse will be ready for embryo transfer to the uterus. (c) The fat pad, ovary, oviduct, and a small section of the uterus have been stabilized by anchoring them in a notch cut in the side of a square of card. Embryo transfer to the oviduct can now be done.

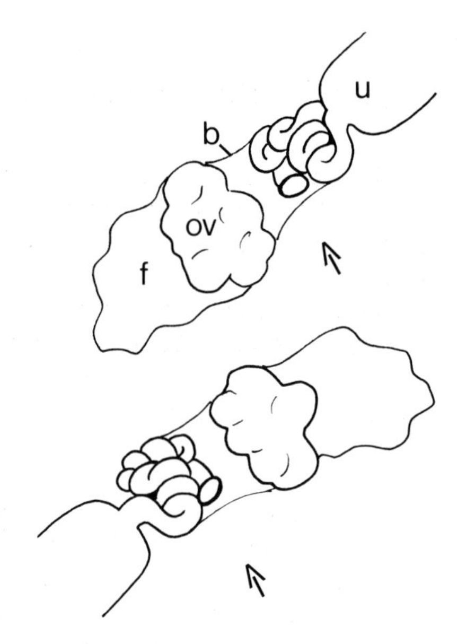

Figure 15. Diagram of the fat pad (f), ovary (ov) coiled oviduct, and distal end of the uterus (u) indicating the position of the ostium when the mouse is prepared as illustrated in *Figure 14*. The embryo transfer pipette should be inserted in the direction of the *arrows* once a hole has been opened in the bursa (b). The right ovary and oviduct is shown at the *top*, the left at the *bottom*.

Protocol 8. Embryo transfer to the uterus

1. Follow steps 1–5 of *Protocol 7* with the modification that a length of uterus is also exteriorized and directed away from the midline.
2. With the loaded embryo transfer pipette and a sewing needle (sharp) or a hypodermic needle in the right hand, and the uterus held firmly by the mesentery with watchmaker's forceps in the left hand, insert the needle through the muscle layers of the uterus, 2 mm from the utero-tubal junction. Direct the needle parallel to the long axis of the uterus so that the tip enters the lumen.
3. Remove the needle and without taking your eyes off the hole, insert the tip of the transfer pipette into the lumen and expel the embryos and marker air bubbles. These will not be visible through the uterine wall but will serve as indicators of the expulsion of the embryos.
4. Check the transfer pipette to ensure that all embryos have been expelled.
5. Replace the tract and repeat on the left side for a bilateral transfer.
6. Close the incision as for oviduct transfers.

---

5. Chimerism

5.1 Detection and quantification of chimerism

5.1.1 Coat colour

The most convenient and readily apparent genetic marker of chimerism is coat colour. The common coat colour alleles at variance in ES cell injection experiments are shown in *Tables 1* and *3*, along with the cell types affected. Chimeric combinations of strains which differ at only one coat colour locus allow a simple visual appreciation of the degree of tissue contribution of each component in terms of the proportion of the coat that expresses the ES cell allele. This evaluation of chimeric animals is necessarily subjective but in general, the degree of coat colour chimerism of a particular animal correlates with the degree of germline contribution. In evaluating a given ES cell clone, its contribution to a number of chimeras should be considered. For example, if a clone consistently contributes more than 50–60% of the coat in a series of chimeras, then it will quite likely contribute to the germline at a similar level. Conversely, if it rarely contributes more than 5–10% of the coat to any chimera, the likelihood of germline transmission is correspondingly small. The decision of whether or not to test-breed a particular chimera should thus depend on its level of chimerism as well as the behaviour of that ES cell clone in other chimeras.

5.1.2 GPI

Another method of detecting chimerism takes advantage of isozyme differences between strains, most commonly in the ubiquitous, dimeric enzyme, glucose phosphosphate isomerase (GPI) which has three variants (*Table 1*) distinguishable by electrophoretic mobilities. Homozygous animals show a single, homodimer band while heterozygotes for any two alleles will have

Table 3. Some common coat colour alleles used as markers in chimera experiments and the cell type they affect.

Locus	Alleles	Cell type affected and phenotypic effect
Agouti	A^w, A, a, a^e	Affects hair follicle function, alleles differ in the amount and distribution of yellow pigment; wild-type (A^w or A) hair has a yellow sub-apical band.
Brown	B, b	Affects melanocytes; wild-type is black (B).
Albino	c, c^{ch}	Affects melanocytes; homozygous albino (c/c) will mask any other coat colour allele; c^{ch} causes a reduced pigmentation and affects yellow pigment more than black; c/c^{ch} is intermediate between c/c and c^{ch}/c^{ch}.
Pink-eyed dilution	p	Affects melanocytes; causes a reduced pigmentation but affects black pigment more than yellow.

both homodimers and the corresponding heterodimer which runs intermediate to the corresponding homodimers. Since dimerization takes place intracellularly, chimeras between homozygotes for different alleles will show only homodimers, in a proportion that reflects the chimeric contributions of each cell type. The only exception is chimeric muscle which may show a heterodimer if the two cell types fuse during myogenesis.

The GPI electrophoretic assay is particularly useful since it can be used on any tissue at any stage of development and will provide a crude measure of the degree of ES cell contribution. About 5% contribution can be detected and the relative strength of bands indicates the proportion of tissue contribution. The enzyme is very hardy; even tissue from animals that have been dead for some hours can be typed successfully and samples can be stored frozen indefinitely. The one disadvantage is that the assay uses tissue homogenates, and so provides no information about the cell types derived from the ES cells. GPI isozymes can be separated by starch gel (*Protocol 9*) or cellulose acetate electrophoresis (13; Chapter 5, *Protocol 9*).

Protocol 9. Horizontal starch gel electrophoresis for GPI isozymes

1. Prepare tissue by adding a drop of distilled water to a small fragment and freezing and thawing several times to lyse the cells. Blood samples need only be frozen and thawed without dilution.

2. To prepare the gel, mix the following ingredients in a 500 ml vacuum flask:
- 12–14 g Electrostarch (Otto Hiller Co.)
- 95 ml dH_2O
- 5 ml Tris–citric acid running buffer (see below)

Heat and stir until the mixture passes through a viscous stage and starts to bubble. Cook 60 sec more, then de-gas under strong vacuum for 45 sec. Pour into three gel moulds to make 8 cm × 10 cm × 5 mm gels and refrigerate for 12–48 h.

3. Remove the gel from the mould and place it on a glass plate. Use a gel comb to cut a row of 5 mm slots across the long side of the gel near one edge. A second row can be put near the middle of the gel.

4. Load the samples on to the gel as follows.

(a) Briefly centrifuge the sample to sediment tissue fragments if necessary.

(b) Cut 2 × 5 mm strips of cellulose acetate that will fit into the slots of the gel.

(c) Using forceps, soak a strip in each sample, blot off excess, and insert into the gel slots, rinsing the forceps between each sample.

5. Run the gel (anode to cathode) in a horizontal electrophoresis chamber, using filter paper wicks, at 120–150 V for 2–3 h using the following Tris/citric acid running buffer, pH 6.5.

- 27 g Tris/litre
- 16 g citric acid buffer/litre

6. At the conclusion of the run, prepare the staining solution immediately before use as follows.

(a) To stain one gel, mix together:

- 5 ml 0.3 M Tris–HCl buffer (36.3 g Tris/litre, adjusted to pH 8 with HCl)
- 1 ml fructose-6-phosphate (20 mg/ml)
- 0.1 ml NADP (10 mg/ml)
- 5 μl glucose-6-phosphate dehydrogenase
- 0.1 ml MTT (3(4,5-dimethyl-2-thiazolyl)-2,5-diphenyl-2H-tetrazoluium bromide 10 mg/ml)
- 0.1 ml phenazine methosulfate (10 mg/ml)

(b) Add this mix to 6 ml of melted 2% agar.

7. When the solution has cooled slightly, pour evenly over the gel and incubate in the dark at 37°C until the colour develops. Under these conditions GPI-1AA runs the slowest, GPI-1CC runs the fastest, with GPI-1BB intermediate.

5.2 Phenotypic effects of chimerism

5.2.1 Sex ratio of chimeras

All of the ES cell lines in common use were derived from male embryos and thus contain a Y chromosome. There are several reasons for preferring to use male ES cell lines: male ES cells produce a higher proportion of phenotypic male chimeras, male chimeras can be bred more rapidly to test for germline transmission, and female-derived XX cell lines are reputed to be unstable. This reputation for instability comes partly from work with female embryonal carcinoma cell lines, which were shown to lose one X after several passages to become XO. Female ES cells can be derived as readily as male ES cells, and there are no clear-cut reasons other than those mentioned above to avoid their use.

A sex bias among ES cell chimeras in favour of males is a common observation when male-derived, XY ES cells are used. In combination with a female embryo, male cells will often produce a fertile, phenotypic male chimera. This sex conversion presumably occurs when XY cells colonize sufficient portions of the various tissues which determine sex in the developing embryo. Sex conversion is not always complete in XX⟷XY chimeras, however, and occasionally results in an infertile hermaphrodite or gynandromorph. Hermaphroditism may be subtle, with the animal showing small testes and abnormalities of the urogenital system, or more obvious with the formation of ovotestes and components of male and female reproductive tracts in the same animal. Such anatomic curiosities are rarely fertile and should be discarded if germline transmission is the object. An exploratory laparotomy will quickly determine the status of an animal suspected of being a hermaphrodite either because of infertility or ambiguous external genitalia.

Fully sex-converted chimeras are advantageous for breeding since they will only transmit the ES cell (XY) genotype, due to the inability of XX cells to undergo spermatogenesis. On the other hand, transmission of the male-derived ES cell genotype has been reported to occur from female chimeras by a number of laboratories. Since XY cells do not normally undergo oogenesis in chimeras, this phenomena may be due to the loss of part or all of the Y chromosome, resulting in effectively XO cells which are capable of forming ova, as was shown in at least one case of transmission from a female chimera (33).

5.2.2 X-linked mutations and heterozygous effects

Chimeras with targeted ES cells contain heterozygous cells; the effects of heterozygosity may reveal themselves to a degree corresponding to the amount of ES cell contribution. In the case of targeted, X-linked genes in male ES cells, chimeric mice are hemizygous for the mutated allele and will demonstrate whatever phenotype is associated with the mutation relative to the amount and type of chimeric tissue. For example, a mutation at the X-linked gene GATA-1, an erythroid-specific transcription factor, causes the death of those chimeras with the greatest contribution of mutant cells, due to failure of erythropoiesis (34). Similarly, lethality associated with increasing amounts of chimerism may arise in the case of an autosomal dominant mutation. These possibilities should be considered when designing targeting experiments, especially since they can define the end-point in cases where the chimeric phenotype is so severe as to prevent germline transmission.

In situations like this, a 'rescue' vector containing a functional copy of the targeted gene, integrated into another chromosome, may help the targeted allele through gametogenesis in the chimera. Segregation will then separate the mutant from the rescue vector insertion in subsequent generations. This will allow the mutant phenotype to be studied in a non-chimera, even if the effect is severe.

6. Maintaining a targeted mutation

6.1 Animal husbandry

Once chimeras have been produced, and throughout subsequent studies of the mutant phenotype, efficient breeding is crucial for an expedient analysis of the mutation. In addition to the time required to produce the targeted ES cell clone, the minimum time between injection of targeted cells into embryos and the birth of mice bred to homozygosity for the mutant allele is five months. Any deficiencies in the breeding programme will increase this time considerably. Thus, the highest standards of animal husbandry must be maintained and the breeding programme of chimeric mice and their progeny must be scrupulously monitored. Monitoring for the presence of murine pathogens should be a routine feature of the health surveillance of the facility, since the presence of any pathogen has the potential to interfere with breeding or the analysis of mutant phenotypes. Conventional mouse facilities can be maintained in such a way as to minimize the possibility of such common pathogens as murine hepatitis virus (MHV) and Sendai virus. The routine use of micro-isolator cages and Hepe-filtered cage changing stations, along with strictly controlled import of animals, with judicious use of quarantine rooms, can keep a conventional facility virtually problem-free. As a final safeguard, cell lines obtained from outside sources can be MAP (murine antibody production) tested to ensure that they do not harbour pathogenic viruses.

6.2 Test-breeding

Chimeric males (and sometimes females) are generally test-bred to ascertain contribution of the ES cells to the germline. It is possible that other methods of determining germline contribution could be developed, such as molecular analysis of sperm DNA using PCR to detect the altered allele. Reasons for screening chimeras in this way, such as suspected effects of the mutation on gametogenesis in the chimera, may make the effort worthwhile, but for the most part, test-breeding should reveal fairly quickly what secrets the germ cells conceal.

Chimeras should be mated to mice with genetic markers that will allow a distinction to be made between ES-derived and host blastocyst-derived gametes, keeping in mind that only half of the ES cell-derived gametes will carry the targeted allele. For example, a chimera between a 129/SvJ-derived ES cell and a non-albino blastocyst could be test mated with any albino mouse. Offspring derived from the host blastocyst gametes would have black eyes, which are distinguishable at birth, while ES-derived gametes would produce progeny with pink eyes. We have used several different 129-derived ES cell lines, differing in coat colour and GPI alleles (*Table 1*), in combination with C57BL/6J blastocysts to produce chimeras (*Figure 16*). A good test-breeding scheme for this common combination is to breed chimeras to C57BL/6J mice. GPI typing can be done at birth or animals can be distinguished on the basis of their agouti phenotype when yellow-banded hairs are evident at about seven days post-natally. Agouti animals are derived from the 129 ES cells and will be (129 × C57)F_1; half of them will carry the targeted allele, provided there is not a dominant lethal effect of the mutation prior to birth, or a haploid effect on sperm function. Non-agouti animals will be pure C57BL/6J and can be raised and used for other experiments.

Figure 16. A chimeric mouse made by injecting 129/SvJ ES cells into a C57BL/6J blastocyst. This mouse has albino hairs, agouti hairs, and black hairs. Note the eye pigmentation.

6.3 Breeding schemes to maintain targeted alleles

Once a germline chimera has been identified, the first priority will be to obtain and maintain the targeted allele in living animals. The second priority will be to put the new allele on to a genetic background(s) that will allow its characterization. For many studies, a random or genetically undefined background may be adequate for the initial characterization, but for others, an inbred background may be preferable.

The best breeding scheme to use will depend on the timing and severity of the effect of the mutant gene. Clearly, dominant lethal mutations can not be maintained at all if the lethality occurs before sexual maturity or if it affects gamete function. In this case, the chimera itself will be the only transmitter of the allele and the mutant effects can only be studied in its offspring prior to their death (see Section 5.2.2). Mutants with other dominant or recessive effects can be maintained by standard schemes for maintaining mutant genes (35).

6.3.1 Segregating backgrounds

If test-breeding is done as described above, the first available heterozygous animals are likely to be F_1 mice between two inbred strains (a 129 substrain and C57BL/6J). If the founder chimera is transmitting the ES cell genotype at a high frequency, it will be possible to produce a large number of F_1 offspring. Once they have been typed by PCR or Southern blot analysis of tail DNA, they can be mated together to produce litters with homozygous normal, heterozygous, and homozygous mutant animals. This breeding scheme takes advantage of the hybrid vigour of F_1 mice and is the fastest means of producing mutant homozygotes for analysis. The drawback is that the F_2 offspring will be segregating for all alleles that differ between the two parent strains and thus the mutant is being observed against a variable genetic background. The practical effect is that any variation in the mutant phenotype between animals might be attributable to unknown, segregating, modifier genes rather than the mutation alone. None the less, this scheme is very useful for a first characterization of genes with major effects.

It is also worth mentioning that mutations unrelated to the targeted locus may have occurred fortuitously in the ES cells at any time during isolation, culture, or cloning. If this occurs, a distinctive phenotype may be detected, but this will segregate independently from the targeted allele and should be evident as a separate locus in one or two generations for all but the most tightly linked loci. Routine genotyping of animals will quickly distinguish between effects of a fortuitous mutation and effects of the targeted allele.

If the founder chimera is transmitting the targeted allele at a very low frequency, or if it ceases breeding after only a few carrier F_1s are produced, the mutant allele can be propagated by crossing heterozygotes to F_1 or even random-bred mice to ensure its survival. Transfer to an inbred background can then be done at any time independently of the founder chimera.

6.3.2 Inbred backgrounds

A uniform genetic background allows the most precise comparison to be made between mutant and non-mutant phenotypes. There are two ways to put targeted alleles on inbred backgrounds: by backcrossing and by mating the chimera with the strain from which the ES cells were derived. The former, which is described below, is straightforward and applicable to any inbred strain, but the latter may be more problematic than it seems if small genetic differences are important. Most commonly used ES cells were derived from 129 mice; however, some were derived many years ago from substrains that are not readily available today, or that might have genetically diverged from the ES cells in the intervening years. An added problem is that some 129 substrains are poor breeders and are thus difficult to maintain. If it is desirable to have the mutant on an inbred background immediately, there may be no choice but to mate the chimera with available 129 mice and to accept the slight genetic variability that will be the inevitable result of crossing two substrains with genetic differences. In test-breeding a chimera with 129 mice, all the offspring, whether derived from the host blastocyst or from the ES cells, will be agouti so that markers other than agouti must be used to distinguish germline transmission. The advantage of this method is that the mutant gene will be immediately available for analysis on a relatively uniform genetic background.

Transferring the mutant allele to another inbred background is a matter of backcrossing. Heterozygous mice are identified by DNA analysis and crossed to mice of an inbred strain. Heterozygous progeny from this mating are then backcrossed to mice of the inbred strain. The more times this cycle is repeated, the more uniform the genetic background becomes. After seven or eight generations, 99% of loci not linked to the mutant allele will be homozygous, and the mice are considered congenic after ten generations, counting the first hybrid or F_1 generation as generation one (35). Backcrossing must continue for another 10–12 generations, however, to ensure genetic uniformity for closely linked loci.

In practice, backcrossing can be done easily, with a minimum number of mice. Starting with an F_1 between the mutant-bearing stock and the inbred strain (generation one), we backcross a single F_1 heterozygous male with several inbred females (generation two). The first litter born is typed and one or two male offspring are again mated with inbred females. All mice of the first generation can be discarded as soon as a pregnancy is detected in the next generation. For at least one backcross cycle, a female heterozygote is mated to an inbred male in order to perpetuate the inbred strain Y chromosome, otherwise it is more efficient to use male heterozygotes. As the stock comes closer to co-isogenicity, the breeding characteristics, along with all other traits, will approximate those of the inbred strain. Usually this will mean a gradual diminution of the hybrid vigour of the early backcrosses, and it may be necessary to set-up additional matings to ensure some of them will be fertile. Designation of the newly produced congenic strain should follow the rules for nomenclature as recommended by the Committee on Standardized Genetic Nomenclature (36). However, as yet there are no specific rules for nomenclature of targeted alleles; they should be given symbols as new alleles of existing loci.

Acknowledgements

We wish to thank James Spencer, John Anderson, and Barbara Van Lingen for technical assistance, Gerry Parker for photographic work, Cindy Welch for preparation of the manuscript, and Allan Bradley for permission to use published figures. This work was supported in part by NIH grant HD27295.

References

1. Tarkowski, A. K. (1961). *Nature,* **190,** 857.
2. McLaren, A. (1976). *Mammalian chimaeras.* Cambridge University Press, Cambridge.
3. Moustafa, L. A. and Brinster, R. L. (1972). *J. Exp. Zool.,* **181,** 193.
4. Brinster, R. L. (1974). *J. Exp. Med.,* **140,** 1049.
5. Papaioannou, V. E., McBurney, M. W., Gardner, R. L., and Evans, M. J. (1975). *Nature,* **258,** 70.
6. Mintz, B. and Illmensee, K. (1975). *Proc. Natl Acad. Sci. U.S.A.,* **72,** 3585.
7. Papaioannou, V. E. and Rossant, J. (1983). *Cancer Surveys,* **2,** 165.
8. Evans, M. J. and Kaufman, M. H. (1981). *Nature,* **292,** 154.
9. Martin, G. (1981). *Proc. Natl Acad. Sci. U.S.A.,* **78,** 7634.

10. Robertson, E. J. (1991). *Biol. Reprod.,* **44,** 238.
11. Gardner, R. L. (1968). *Nature,* **220,** 596.
12. Papaioannou, V. E. (1981). In *Techniques in the life sciences* Vol. P1/1, *Techniques in cellular physiology part 1* (ed. P. F. Baker), pp. 116/1–27. Elsevier/North Holland Biomedical Press, Amsterdam.
13. Bradley, A. (1987). In *Teratocarcinomas and embryonic stem cells: a practical approach* (ed. E. J. Robertson), pp. 113–51. IRL Press, Oxford.
14. Hetherington, C. M. (1987). In *Mammalian development: a practical approach* (ed. M. Monk), pp. 1–12. IRL Press, Oxford.
15. Stevens, L. C. (1983). In *Teratocarcinoma stem cells, Cold Spring Harbor Conferences on Cell Proliferation*, Vol. 10 (ed. L. M. Silver, G. R. Martin, and S. Strickland), pp. 23–36. Cold Spring Harbor Laboratory.
16. Axelrod, H. R. (1984). *Develop. Biol.,* **101,** 225.
17. Robertson, E. J., Kaufman, M. H., Bradley, A., and Evans, M. J. (1983). In *Teratocarcinoma stem cells, Cold Spring Harbor Conferences on Cell Proliferation,* Vol. 10 (ed. L. M. Silver, G. R. Martin, and S. Strickland), pp. 647–64. Cold Spring Harbor Laboratory.
18. Hay, R. J., Macy, M. L., and Chen, T. R. (1989). *Nature,* **339,** 487.
19. Johnson, R. S., Speigelman, B. M., and Papaioannou, V. E. (1992). *Cell,* **71,** 577.
20. Doetschman, T. C., Eistetter, H., Katz, M., Schmidt, W., and Kemler, R. (1985). *J. Embryol. Exp. Morph.,* **87,** 27.
21. McMahon, A. P. and Bradley, A. (1990). *Cell,* **62,** 1073.
22. Bradley, A., Evans, M., Kaufman, M. H., and Robertson, E. (1984). *Nature,* **309,** 255.
23. Robertson, E., Bradley, A., Kuehn, M., and Evans, M. (1986). *Nature,* **323,** 445.
24. Nagy, A., Rossant, J., Nagy, R., Abramow-Newerly, W., and Roder, J. (1993). *Proc. Natl Acad. Sci. U.S.A.,* in press.
25. Handyside, A. H., O'Neill, G. T., Jones, M., and Hooper, M. L. (1989). *Roux. Arch. Devel. Biol.,* **198,** 48.
26. Champlin, A. K., Dorr, D. L., and Gates, A. H. (1973). *Biol. Reprod.,* **8,** 491.
27. Rugh, R. (1990). *The mouse: its reproduction and development.* Oxford Science Publications.
28. Hogan, B., Costantini, F., and Lacy, E. (1986). *Manipulating the mouse embryo.* Cold Spring Harbor Laboratory.
29. Papaioannou, V. E. and Fox, J. G. (1993). *Laboratory Animal Science* (in press).
30. Papaioannou, V. E. (1990). In *The post-implantation mammalian embryo: a practical approach* (ed. A. J. Copp and D. Cockcroft), pp. 61–80. IRL Press, Oxford.
31. Babinet, C. (1980). *Exp. Cell. Res.,* **130,** 15.
32. Schmidt, G. and O'Sullivan, J. F. (1987). *Trends Genet.,* **3,** 332.
33. Kuehn, M. R., Bradley, A., Robertson, E. J., and Evans, M. J. (1987). *Nature,* **326,** 295.
34. Pevny, L., Simon, M. C., Robertson, E., Klein, W. H., Tsai, S.-F., D'Agati, V., Orkin, S. H., and Costantini, F. (1991). *Nature,* **349,** 257.
35. Green, E. L. (1966). In *Biology of the laboratory mouse* (ed. E. L. Green), pp. 11–22. McGraw-Hill Book Co., NY.
36. Lyon, M. F. and Searle, A. G. (ed.) (1990). *Genetic variants and strains of the laboratory mouse.* Oxford University Press.

Appendix

Further details and suppliers of specialized equipment used in the authors' laboratories. These are not the only possible brands of equipment or suppliers. Requirements and additional information can be found in the text.

Adrenalin chloride solution: Parke Davis, Warner Lambert, Morris Plains, NJ, Cat. No. N0071-4011013.
Defonbrune microforge: Arenberg Sage Inc., 57 Cornwall St, Jamaica Plain, MA 02130.
Dental spatula for plug checking: Schein JE SPT 313, Dental Supply Co. of New England, 80 Fargo St., Boston, MA 02210.
Electrostarch for GPI-1 electrophoresis: Otto Hiller Co., PO Box 1294, Madison, WI 53701.
Kopf Pipette Puller Model 700C: David Kopf Instruments, 7324 Elmo St., PO Box 636, Tijunga, CA 91042–0636.
Microplasma TC Detection Kit: Cat. No. 1004, Gen-Probe, San Diego, CA 92123.

Microscopes
Dissecting scope for embryo recovery: Wild M5A, Wild Hierbrugg Ltd or Parco SM series, Parco Scientific Co., Instrument Division, 316 Youngstown-Kingsville Rd., PO Box 189, Vienna, Ohio 44473.
Dissecting scope for embryo transfer: Reichart-Jung, Stereo Star Zoom (formerly AO Stereostar Zoom) model 570, 0.7 × to 4.2 ×, 15 × widefield eyepieces.
Microscope for embryo injection: Zeiss Standard 14 microscope or Zeiss IM inverted, Carl Zeiss.
Stage cooling device: Physitemp TS-4 Controller, Sensortek Inc., 154 Huron Ave., Clifton, NJ 07013.

Syringe devices for cell injection
Beaudouin suction-and-force pump: 1 et 3 rue Rataud, Paris, 5^{e} France.
Stoelting Co. micrometre with Hamilton syringe: 1350 So, Kostner Ave, Chicago, Il 60623.

Syringe device for holding embryo: Narishige model No. IM-58
Surgical instruments: Roboz Surgical Instrument Co., Inc., 9210 Corporate Blvd., Ste. 220, Rockville, MD 20850.
Watchglasses: Carolina Biological, # 74-2300/2700 York Rd, Burlington, NC 27215.
Wound clips: Autoclip 9 mm, Clay Adams, Division of Becton Dickinson & Co., Parsippany, NJ 07064.

5

Production of completely ES cell-derived fetuses

ANDRAS NAGY and JANET ROSSANT

1. Introduction

Embryonic stem (ES) cells behave like normal embryonic cells if they are returned to the embryonic environment by injection into a host blastocyst or aggregation with blastomere stage embryos. They can contribute to many kinds of tissues in the resulting chimeras suggesting that ES cells have the full potential to develop along all lineages of the embryo proper. Their ability to contribute to extra-embryonic lineages, however, is more restricted (1, 2). For many purposes, it would be advantageous to generate mice that are entirely ES cell-derived, rather than mice which contain both ES and embryonic cell contributions. The fact that ES cells retain a wide developmental potential suggested that this might be possible if ES cells were combined with developmentally compromised embryonic cells. Tetraploid mouse embryos appeared to be good candidates as ES cell partners for such experiments. On their own, tetraploid embryos develop very poorly post-implantation and rarely make it as far as midgestation (3). In chimeras with normal (diploid) embryos tetraploids rarely contribute to the embryo proper. However, their contribution to the extra-embryonic membranes can be relatively extensive (4).

Recently, we have shown that the opposite developmental capabilities of ES cells and tetraploid embryos can indeed complement each other in aggregation chimeras. This complementation resulted in the formation of polarized chimeras in which the fetuses were ES cell-derived and most of the extra-embryonic tissues were provided by the tetraploid component (2).

The aggregation method for making ES cell chimeras (5, 6) offers the advantage over the technique of injecting ES cells into host blastocysts (described in detail in Chapter 4), that it does not require expensive instrumentation and sophisticated manual skills. Many aspects of the production of ES cell–tetraploid embryo and ES cells–diploid embryo chimeras are similar, and so we describe the general procedures for making ES cell aggregation chimeras, then in addition, the production of tetraploid embryos and the few special features of ES cell–tetraploid chimeras. We also briefly describe genetic markers that can be used to analyse mosaicism in the chimera.

To date, we have found only one ES cell line (R1, (7)) which can give rise to viable completely ES cell-derived animals at its early passage number. The completely ES cell-derived fetuses from other cell lines, (e.g. D3(8), AB1(9), α(2)) have not survived beyond birth. The production of such fetuses, however, has various applications. We describe here some of these and provide a protocol for the repopulation of the adult haemopoietic system with ES cell-derived fetal liver cells.

There are great differences among individual ES cell lines and even among sublines of a given ES cell line in terms of their pluripotentiality. This variability is most clearly manifested when the cells are challenged to form the entire embryo. ES cell lines with good developmental potential should support complete fetal development to term at a frequency of about 10–15% of ES cell–tetraploid embryo aggregates transferred to the uterus.

2. Embryonic stem cells

2.1 Preparation for aggregation

ES cell lines to be used to form aggregation chimeras should be maintained under the optimal conditions for the particular line used (see Chapter 2). Cells should be passed at least once after thawing, before using them for aggregation. The protocol given here was developed for R1 (7), and D3 cells (8), and our laboratory conditions. The goal is to produce clumps of 15–25 loosely connected cells just before the aggregation is set-up.

Protocol 1. Preparation of ES cells for aggregation chimeras

Materials

- trypsin (Chapter 2, Section 2.2)
- ES cell medium (Chapter 2, Section 2.2)
- tissue culture plates

Method

Four days prior to aggregation (day −4)

1. Thaw ES cells or passage a confluent plate of cells as described (Chapter 2, *Protocol 4*) and dilute cells one in ten.

Two days prior to aggregation (day −2)

2. Trypsinize a subconfluent plate (60 mm) of ES cells to give a single cell suspension as described in Chapter 2, *Protocol 4*. After spinning down and resuspending the cells in 5 ml medium, let any large clumps settle in the tube for 5 min.

3. Plate approximately one sixth of the cell suspension on a 60 mm plate containing embryonic fibroblast feeders (Chapter 2, *Protocol 2*).

Day of aggregation (day 0), (after preparation of the embryos, see *Protocol 9*)

4. Quickly rinse the plate of ES cells with 3 ml of trypsin solution, then replace with 0.5 ml of fresh trypsin.
5. Trypsinize cells for 5 min at 37°C in 5% CO_2, until they form clumps of loosely connected cells (see *Figure 5B*), then add 5 ml of ES cell medium.
6. Keep the cell suspension in the plate and use to assemble aggregates within the next hour.

3. Production of embryo partners

The importance of the host embryo genotype depends on the type of aggregates. Under optimal *in vitro* culture conditions we found no difference between the development of ES cell–tetraploid embryo aggregates produced with different host genotypes. However, for germline transmission through diploid chimeras, the host genotype might be as important as it is in blastocyst injection (see Chapter 4).

Embryos should be recovered in M2 medium (10) but maintained in M16 (11) at 37°C and 5% CO_2 in air (*Protocol 2*). Since the aggregates are cultured at least for 24 hours, the quality of culture conditions is more critical than for injection chimeras, where the embryos are exposed to *in vitro* conditions for only a few hours. Of the conditions the most critical is the culture media. General notes on media preparation are the following.

(a) Make up all media and stock solution with good quality, at least twice glass distilled or Q-water.
(b) Use plasticware if it is possible or the glassware should have been used for tissue culture only.
(c) Media should be made up fresh every other week.

Protocol 2. Preparation of M2 and M16 from stock solutions

Stock A (10 ×)	g/ml	Source	Cat. No.
NaCl	5.534	BDH	B-10241-34
KCl	0.356	BDH	B-10198-34
KH_2PO_4	0.162	BDH	B-10203-34
$MgSO_4{\cdot}7H_2O$	0.293	Fisher	M-63-500
sodium lactate	2.610	Fisher	S-326-500
or 4.349 g of 60% syrup			

Protocol 2. *Continued*

Stock A (10 ×)	g/ml	Source	Cat. No.
glucose	1.000	Sigma	G-8270
penicillin	0.060	Sigma	Pen-Na
streptomycin	0.050	Sigma	S-6501

store in 10 ml aliquots at 4°C

Stock D (100 ×)			
$CaCl_2 \cdot 2H_2O$	0.252	BDH	B-10070-34

Store in 1 ml aliquots at −20°C

M16

1. Weigh out 3.6 mg pyruvate (Sigma, P-2256), 210.1 mg $NaHCO_3$ (BDH, E-10247-34), and dissolve in 80 ml 2 × distilled or Q-water in plastic beaker.
2. Add 10 ml of stock A and 1 ml of stock D.
3. Sprinkle 400 mg BSA (Sigma, A-4378) on top of the medium and allow to dissolve slowly (10 min.).
4. Make up to 100 ml with 2 × distilled or Q-water.
5. Filter sterilize into tissue culture bottles (two 50 ml bottles).
6. Equilibrate the medium with CO_2 by keeping it (loose caps) in the incubator overnight.
7. Close caps and store M16 at 4°C.

M2

8. Weigh out 3.6 mg pyruvate, 34.9 mg $NaHCO_3$, 496.9 mg Hepes (GIBCO, 845-1344 IM), and dissolve in 80 ml 2 × distilled or Q-water in plastic breaker.
9. Add 10 ml of stock A and 1 ml of stock D.
10. Sprinkle 400 mg BSA on top of the medium and allow to dissolve slowly (10 min.).
11. Adjust the pH to 7.4 with 0.2 M NaOH.
12. Make up to 100 ml with 2 × distilled or Q-water.
13. Filter sterilize into tissue culture bottles (two 50 ml bottles).
14. Store M2 at 4°C.

3.1 Recovery of eight- and two-cell stage embryos

The only point of note is that for aggregation chimeras non-compact eight-cell stage embryos are preferred. Flushing should therefore be performed in the morning of the day of aggregation.

Protocol 3. Flushing of eight-cell stage embryos

Materials

- dissecting microscope
- No. 30 gauge needle (first, the sharp tip is cut off by a strong pair of scissors and then the newly blunted tip is further rounded using a kitchen knife sharpening stone)
- 1 ml syringe
- No. 5 forceps (Dumont)
- dissecting instruments (fine-pointed scissors, fine forceps)
- mouth pipette
- Pasteur pipettes
- alcohol or Bunsen burner
- sterile plastic Petri or tissue culture dishes
- organ culture dishes (Falcon, 3037)
- M2 and M16 medium

Method

1. Kill the superovulated (see *Protocol 4*) or naturally mated females by cervical dislocation on day 2.5 (day 0.5 is the day of the plug).
2. Open the body cavity by making a small ventral incision in the skin across the midline at the level of the bladder, then pull the skin up over the chest and cut the body wall along both sides starting from the midline at the bladder.
3. Dissect the oviducts with the upper part of the uterus attached and place in drop of M2 in a Petri dish (see Chapter 4, *Protocol 4* for details).
4. Transfer one of the oviducts into a small drop of M2.
5. Under dissecting microscope locate the fimbrial end of the oviduct and insert the needle attached to a 1 ml syringe filled with M2. Hold the needle in position with the No. 5 forceps and flush ~0.05 ml of medium through the oviduct (you should see the oviduct swell).
6. Repeat steps 4 and 5 with the remaining oviducts.
7. Collect embryos with a mouth pipette attached to a drawn-out Pasteur pipette.
8. Wash the embryos through several M2 drops to get rid of the debris.
9. Wash them in M16, transfer them into an organ culture dish containing M16, and place them into the incubator (at 37°C, 5% CO_2 in air).

Recovery of two-cell stage embryos which are used to produce tetraploid embryos is very similar to that of the eight-cell stage embryos (*Protocol 3*) and it must be done a day before assembling the aggregates. Many mouse strains exhibit the so-called two-cell stage block. This means that their embryos have difficulties in developing through the two-cell stage *in vitro*. This problem can be overcome in optimal culture conditions but in order to avoid potential problems, it is safer to recover late two-cell stage embryos. The presence of 10–15% three/four-cell stage embryos among two-cell stages at the time of flushing indicates appropriate timing. If the females are superovulated, the recommended time for flushing is 44–46 hours post-hCG for most strains and lines.

For a maximum yield of embryos after superovulation, young females (four weeks of age) are most favourable. However, oviduct flushing of such young females is difficult, since the utero-tubal junction is usually very narrow. The majority of two-cell stage embryos can be damaged unless the junction is ruptured with fine forceps before flushing.

Protocol 4. Production of two-cell stage embryos for blastomere electrofusion

Materials

- pregnant mare serum gonadotropin (PMSG)
- human chorionic gonadotropin (hCG)
- see the list in *Protocol 3*

Method

Day 0 is the day of assembling aggregates.

1. Day −5. Begin superovulation of young (four weeks of age is preferred) females by injecting 5 IU PMSG intraperitoneally between 2–4 pm.
2. Day −3. Give 5 IU hCG at 1 pm and set-up the females for mating with stud males.
3. Day −2. Check plugs.
4. Day −1. Start dissecting oviducts at 10 am and flush oviducts as described for eight-cell stages (see *Protocol 3*). Before flushing, however, rupture the utero-tubal junction by tearing with two No. 5 forceps. Avoid long exposure to room temperature.

3.2 Production of tetraploid embryos

Several different methods have been used to produce tetraploid mouse embryos. These methods can be classified into two categories. One group of techniques utilizes the effect of agents like colchicine (12) or Cytochalasin B (13), that can prevent cytokinesis without influencing nuclear division. The techniques in the second category are based on the induction of fusion of two normal diploid embryonic cells, such as the blastomeres of two-cell stage mouse embryos. Most of the procedures in this category suffer from variation in efficiency caused by the varying fusogenic activity of the agents used such as Sendai virus (14) and polyethylene glycol (PEG) (15). Electrofusion as devised by Kubiak and Tarkowski (16), is very simple and efficient, making it superior to other techniques. Here we describe this method in detail.

Fusion of the blastomeres of two-cell stage mouse embryos occurs when a square pulse of 1 kV/cm field strength, for a duration of 100 μsec is applied perpendicular to the plane of contact of the two cells. The precise parameters of the pulse will vary depending on the equipment used. The pulse shape of the CF-100 pulser, (commercially available from Biochemical Laboratory Service, 31 Zselyi Aladar utca, Budapest, H-1165, Hungary), that we use is a peak with a square shoulder. Therefore our effective field strength (3–3.5 kV/cm) is higher than the minimum required. The equipment was designed specifically to meet the needs of blastomere fusion. The pulse amplitude and duration can be set within the range of 0 to 100 V and 0 to 120 μsec, respectively. The machine also provides an adjustable strength 1 MHz AC field to allow correct orientation of the two-cell stage embryos between the two electrodes in non-electrolyte solution. A hand-held trigger provides a convenient means of initiating the pulse while working with embryos under the microscope. Two kinds of electrode-chambers are provided; electrodes with adjustable distance (they are connected to simple micromanipulators), and electrodes with a fixed distance (250 μm). All the parameters given below are applicable to the above equipment.

3.3 Electrofusion in non-electrolyte solution

In order to fuse the blastomeres the plane of contact between the two cells must be perpendicular to the electric field generated between the electrodes. The main advantage of non-electrolyte fusion is that proper orientation of the embryos is provided by the 1 MHz AC field of the equipment (*Figure 1A*). In electrolyte fusion the embryos must be oriented manually. As a consequence, in the latter case the embryos must be shocked individually, one at a time, whereas in non-electrolyte you can apply the electric pulse to a group of 20–40 embryos.

The effective voltages for your equipment and electrode slide must be determined in a pilot experiment by following the steps of *Protocol 5* with fewer embryos in the groups. Choose the voltage and duration which causes no embryo lysis and still fuses the majority of the two-cell stages. This will be close to 100 V and 20–30 μsec for electrodes placed 250 μm apart. The effective orienting AC field (around 0.5 V) must be precisely determined as well. Too high an AC field also causes immediate lysis.

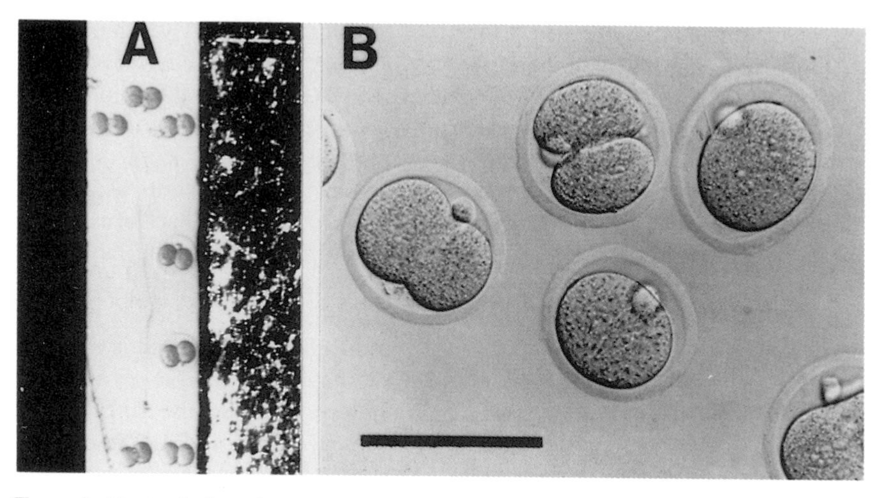

Figure 1. Electrofusion of two-cell stage mouse embryos in non-electrolyte. (A) Embryos under orienting AC field. (B) Embryos during the fusion process. Bar = 100 μm.

Protocol 5. Preparation of the fusion chamber and fusion in non-electrolyte (day −1)

Materials

- 0.3 M mannitol (Sigma M4125) (filtered through 0.22 μm Millipore filter)
- M2 medium (see *Protocol 2*)
- two dissecting microscopes
- M16 microdrops covered by paraffin oil (see *Protocol 7*) in 35 mm tissue culture dishes
- CF-100 pulse generator
- electrode-chamber

Method

1. Put a 100 mm Petri dish containing the electrode-chamber under a dissecting microscope and connect the cables to the pulse generator.
2. Place two large drops of M2 medium (drop A and D), and two drops of mannitol solution (drop B and C) in the dish as shown in *Figure 2*. Drop C should be eccentric with respect to the electrodes, to avoid blocking of vision by accidental air bubbles.
3. Set the pulse generator to the effective DC and zero AC voltages, and the effective (20–30 μsec) pulse duration. Switch the equipment on and make sure that everything is connected properly.

4. Place all the embryos in drop A.
5. Pass 15–20 embryos through drop B and place them in drop C between the electrodes. Slowly increase the AC field. This will properly orient most of the embryos. Pick up those which are not oriented and let them fall back again.
6. When all the embryos are properly oriented push the trigger to apply the fusion pulse.
7. Transfer the embryos into drop D.
8. Transfer the group of 15–20 embryos into a microdrop of M16 under oil (prepared similarly to that described in *Protocol 7*) and put them into the incubator.
9. Repeat steps 5–8 with new groups of embryos.

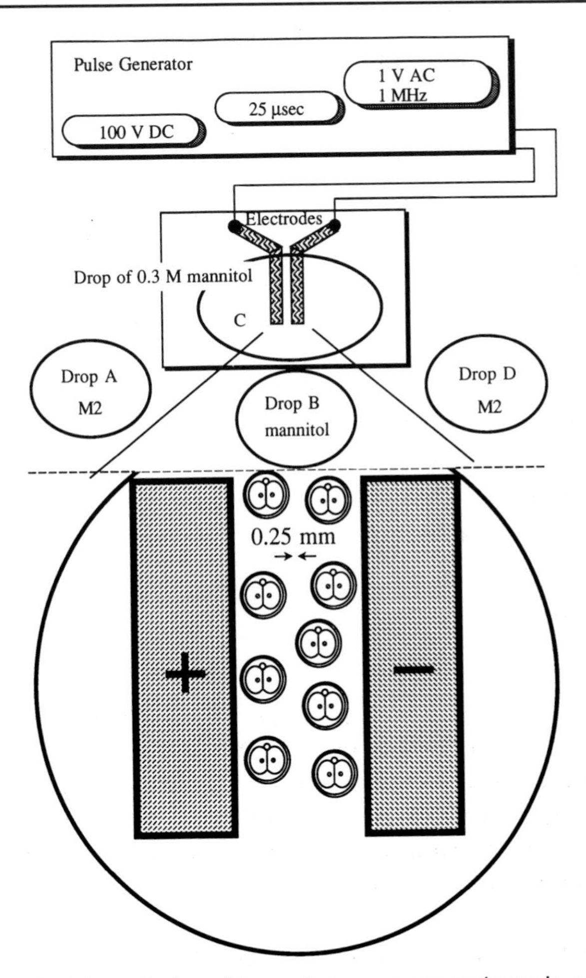

Figure 2. Protocol of electrofusion of two-cell stage mouse embryos in non-electrolyte.

3.4 Electrofusion in electrolyte

Effective fusion can also be achieved if the fusion impulse is applied in electrolyte solution, like PBS or culture medium (M2). Since the automatic embryo orientation provided by the AC field does not work in electrolyte, orientation must be done manually and individually. The set-up is shown in *Figure 3*. The effective DC voltage should be determined in a pilot experiment. By contrast to non-electrolyte fusion, too high voltage does not cause immediate lysis of the embryos, but frequently causes vacuole(s) in the cytoplasm after overnight culture, which may interfere with normal development of tetraploids. Find the lowest voltage which still fuses more than 95% of the embryos.

Protocol 6. Preparation of the fusion chamber and fusion in electrolyte

Materials

- M2 medium (*Protocol 2*)
- two dissecting microscopes
- M16 microdrops covered by paraffin oil (see *Protocol 7*) in two or three 35 mm tissue culture dishes
- CF-100 pulse generator
- electrode-chamber

Method

1. Place a 100 mm Petri dish containing the electrode-slide under a dissecting microscope and connect the cables to the pulse generator.
2. Put a large drop of M2 medium over and eccentric to the electrodes, to avoid blocking of vision by accidental air bubbles.
3. Set the pulse generator to the effective DC voltage (around 100 V), 100 μsec duration and 0 V AC field, switch the equipment on and make sure that everything is connected properly.
4. Transfer 20–40 embryos to compartment A of the drop.
5. Pick up one embryo and place it between the electrodes. Orient properly by blowing on the side of the embryo with medium.
6. Push the trigger to apply the fusion pulse.
7. Pick up the embryo and transfer it into compartment B.
8. Repeat steps 5–7 with new embryos from compartment A.
9. When the group is finished tansfer the embryos back to M16 drops under oil (prepared similarly to that described in *Protocol 7*) and return to the incubator.
10. Repeat the procedure from step 4 with the remaining embryos.

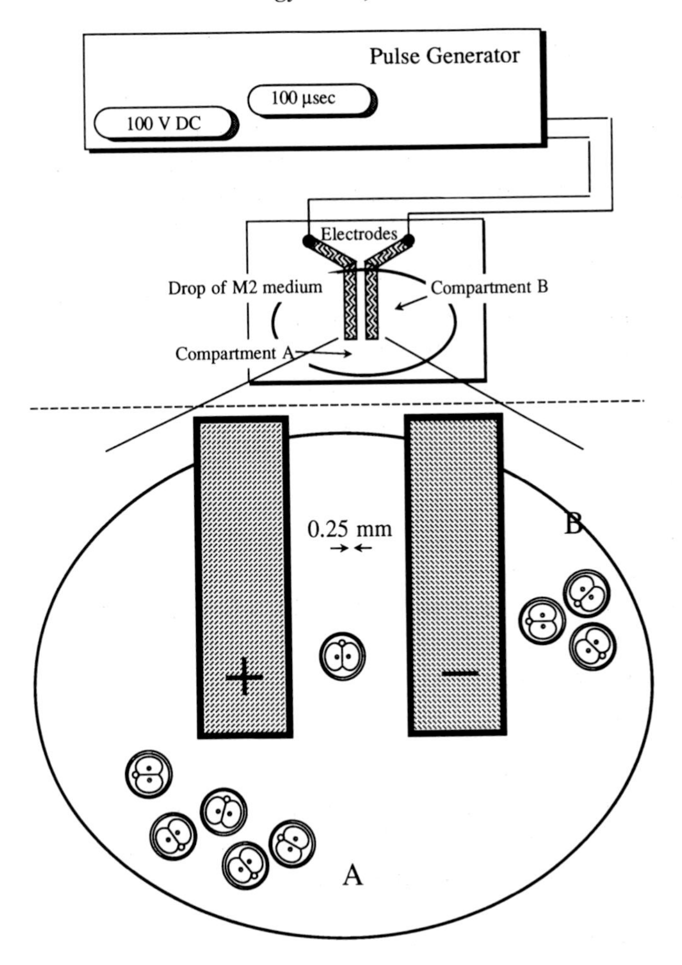

Figure 3. Protocol of electrofusion of two-cell stage mouse embryos in electrolyte.

3.5 Process of fusion, selection, culture

Fusion of the blastomeres is complete in 30 minutes. It occurs at room temperature, but is faster and more effective at 37°C. The boundary between the two cells completely disappears, the embryos round up, and they resemble one-cell stage embryos (see *Figure 1B*).

Since embryos are recovered at the late two-cell stage, the second mitotic division is expected soon after the fusion. Therefore it is important to select for the perfectly fused tetraploid embryos 30–60 minutes after application of the electric pulse. Occasionally one of the blastomeres lyses during fusion; such embryos will only contain one viable cell and could be confused with fused embryos. However, these embryos are readily recognized during the process of lysis right after applying the electric pulse or later by their size, which is half normal. It is safest to transfer the tetraploids into a new culture dish or new M16 microdrop and leave the unfused and lysed embryos at their old place. Under optimal conditions the rate of unfused and lysed embryos does not exceed 5–10%. We place 15–20 embryos into one microdrop of M16 medium for overnight culture.

Electrofusion of pairs of blastomeres of four-cell stage embryos is also possible. Due to the non-planar arrangement of cells, the four-cell stage embryo has to be subjected to two pulses from different directions. In this case the proper orientations can only be done manually.

4. Aggregating ES cells with cleavage stage embryos

4.1 Preparation of aggregation plate

The aggregation plate is where the actual assembly and overnight culture of aggregates takes place. The plate described below holds up to 60 aggregates. It has microdrops for the final selection of the ES clumps (see top and bottom rows of *Figure 4A*) and for the aggregates (middle two rows of *Figure 4A*). In the latter drops six small depressions are made in the plastic with a darning needle that serve as cradles for the aggregates (*Figure 4B*). The walls of the depressions bring the embryos and ES cells into close proximity and therefore promote their aggregation. The type of darning needle used is critical. The tip must have a smooth right angled cone shape and intact, factory-made coating. Check several and choose the best. Never flame the needle, but burn a drop of ethanol at its tip to sterilize it before use. A yellow or white pipetteman-tip

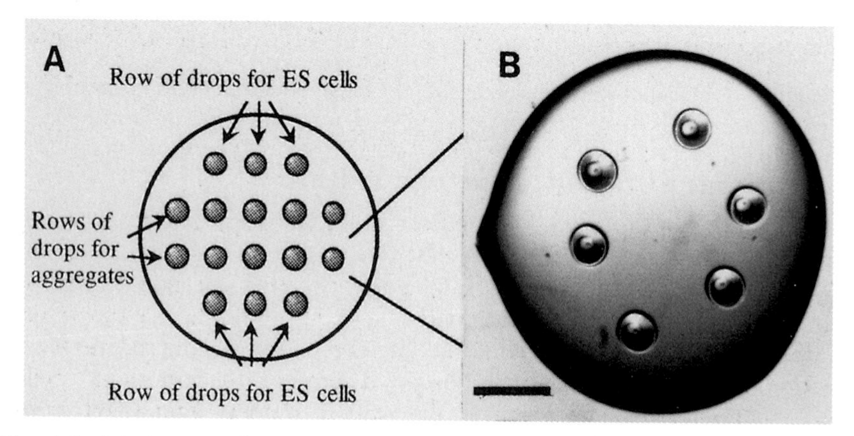

Figure 4. Arrangement of microdrops in aggregation plate (A), and that of the depressions (B), made by darning needle in the microdrops of second and third rows. Bar = 1 mm.

pulled over the eye of the needle gives a strong area for holding. We use darning needles imported from Hungary. If you have a problem finding a source of good needles, we may be able to provide a suitable one on request.

Protocol 7. Preparation of aggregation plate (day 0)

Materials

- 35 mm bacteriological Petri dish
- M16 medium (*Protocol 2*) in 1 ml syringe with 26 G needle
- paraffin oil (Fisher Scientific, Light Mineral Oil, 0121-B)
- darning needle

Method

1. Using the M16 filled syringe, put four rows of M16 microdrops (roughly 3 mm in diameter) into a 35 mm tissue culture dish, three drops in the first and fourth, five drops in the second and third rows (*Figure 4A*).
2. Cover the whole plate with paraffin oil.
3. Sterilize the needle by burning an ethanol drop at its tip. Do it carefully to avoid burning the factory-made coating of the needle. Let the needle cool down for a few seconds.
4. Press the cool darning needle into the plastic through the paraffin oil and culture medium, while making a circular movement with the free end of the needle you are holding. Do not twist! This movement creates a tiny scoop of about 300 μm in diameter with a clear smooth wall.
5. Make six such depressions in each microdrop of the second and third rows (*Figure 4B*).
6. Return the plate to the incubator while you are preparing the embryos.

4.2 Preparation of diploid or tetraploid embryos

4.2.1 Selection of embryos

As far as diploid embryos are concerned the selection is obvious. Choose only perfect eight-cell stage embryos for aggregation.

After fusion the majority of tetraploid embryos (usually about 80–90%) cleave at the time of the normal second mitosis and reform the two-cell stage. By noon of the next day they have cleaved once more and reached the four-cell stage. Since this stage is equivalent to the diploid eight-cell stage, tetraploid embryos start compacting at this stage. The remaining tetraploid embryos (about 10–20%) cleave slower or miss one or more cleavages. These embryos are still at the two-cell stage at noon the next day. Discard the delayed embryos. Our experience shows that they are not able to support the development of ES cells in chimeras.

4.2.2 Zona removal

There are two different ways to remove the zona pellucida of the embryos; one is enzymatic and uses pronase (17), the other is dissolution and uses an acidified saline solution. Both are very effective and do not have any negative effects on the further development of the embryos. The only practical difference is the speed of the reactions. Zona removal by pronase takes about 10–15 minutes, while acid-saline works in about one minute. Therefore, we recommend the use of acid treatment. One potential problem encountered with this approach is that since the acid pH is critical for the reaction, transfer of too much medium along with the embryos to the acid drop can cause the pH to elevate. Conversely, transfer of too much acid solution with the embryos after zona dissolution can cause the pH of the medium to alter. To eliminate these problems always transfer minimal amounts of solutions with the embryos and use multiple washes.

Protocol 8. Removal of zona pellucida by acid Tyrode's solution

Materials

- bacteriological Petri dish
- acid Tyrode's (room temperature or colder)

Components	g/100 ml
NaCl	0.800
KCl	0.020
$CaCl_2 \cdot 2H_2O$	0.024
$MgCl_2 \cdot 6H_2O$	0.010
glucose	0.100
polyvinylpyrrolidone (PVP)	0.400

adjust pH to 2.5 with 5 M HCl then filter sterilize

- M16 medium in 1 ml syringe
- dissecting microscope
- mouth pipette
- Pasteur pipette

Method

1. Put two drops of M16 and several drops of acid Tyrode's in the bacteriological Petri dish.
2. Transfer all of the embryos quickly into one of the M16 drops.

3. Pick up 20–50 embryos (the number depends on how quickly you can handle them) with as little medium as you can, and wash through one drop of acid Tyrode's, then transfer to a fresh drop.
4. Agitate the embryos in the acid solution while observing zona dissolution.
5. Transfer the embryos into the fresh drop of M16 as soon as their zonas dissolve.
6. Repeat the procedure with the remaining embryos. Use new drops of acid Tyrode's for each group of embryos.
7. Wash the embryos through two to three drops of M16 before putting them in the aggregation plate.

4.3 Assembly of aggregation 'sandwiches'

ES cell aggregations are made by 'sandwiching' ES cells between two embryos. The rationale for this is that the cells are readily internalized as the two embryos aggregate. As far as the ES cell component is concerned, we routinely use 10–15 ES cells for each aggregate. This number was determined experimentally. Higher numbers of cells was found to interfere with normal development, and lower cell numbers were hard to manipulate. Loosely connected clumps of 10–15 cells can be handled easily and do not impair development.

Assembling aggregation 'sandwiches' as described in *Protocol 9* does not require as much technical skill as microinjecting cells into blastocysts. In the course of half an hour, with practice you can assemble one hundred aggregates.

Protocol 9. Assembly of 'sandwich-type' ES cell aggregation chimeras

Materials

- prepared aggregation plate (*Protocol 7*)
- dissecting microscope
- mouth pipette
- Pasteur pipette

Method

1. Directly after the removal of the zona pellucida, transfer groups of 12 embryos into each depression-containing microdrop of the aggregation plate. In order to prevent accidental aggregation while you are preparing the ES cells, put one embryo into and one beside each depression (see *Figure 5A*).

Protocol 9. *Continued*

2. Trypsinize ES cells as described in *Protocol 1*.
3. Choose a number of clumps of loosely connected cells (*Figure 5B*) from the plate under the dissecting microscope and transfer them into the first and fourth rows of the aggregation plate (there are no embryos in these microdrops). Make sure that the clumps are not too dense (not more than 80–100 clumps per drop), so you can easily make the final, individual selection among them.
4. Carefully select at least six clumps of ES cells of the required cell number (usually 10–15 cells), transfer them to a drop containing embryos (see *Figure 5C*), and release them at a distance from the embryos.
5. Pick up one of the clumps and let it fall on the side of the embryo sitting in a depression. Pick up the embryo placed outside the depression and place it on the free side of the ES cell clump (see *Figure 5D*).
6. Assemble all the aggregates in this manner, check the plate, and then put the plate back into the incubator for overnight culture.

4.4 Culture and transfer of embryos

Usually more than 90% of the sandwiches aggregate overnight (*Figure 5E*) and form early blastocysts by the afternoon of the next day (*Figure 5F*). At this time they should be transfered into the uterus of pseudopregnant recipients (see Chapter 4, *Protocol 8*). We transfer a maximum of eight embryos into both uterine horns. In a situation where there are too many embryos for the number of recipients you may culture the aggregates for one more day and transfer into two and a half day pseudopregnant recipients at this time.

4.5 Caesarean section

Depending on the nature of the experiment the embryos derived from the aggregates can be recovered at different stages of development. The developmental stage of the embryos is determined by the recipient. Methods of dissection of many embryonic stages are found in Chapter 6, Section 4.2. Therefore here we only describe the recovery of the neonate by Caesarean section. Caesarean section is necessary to recover completely ES cell-derived or extensive ES cell chimeras at the newborn stage, since both of them experience severe survival problems. If we allow the recipients to give birth naturally, these newborns usually die shortly after birth and are eaten by the mothers or are dead long before they are noticed. If you are planning to foster the newborns that survive after Caesarean section, you must have foster mothers available by setting up normal matings that precede the pseudopregnant recipients by one or two days.

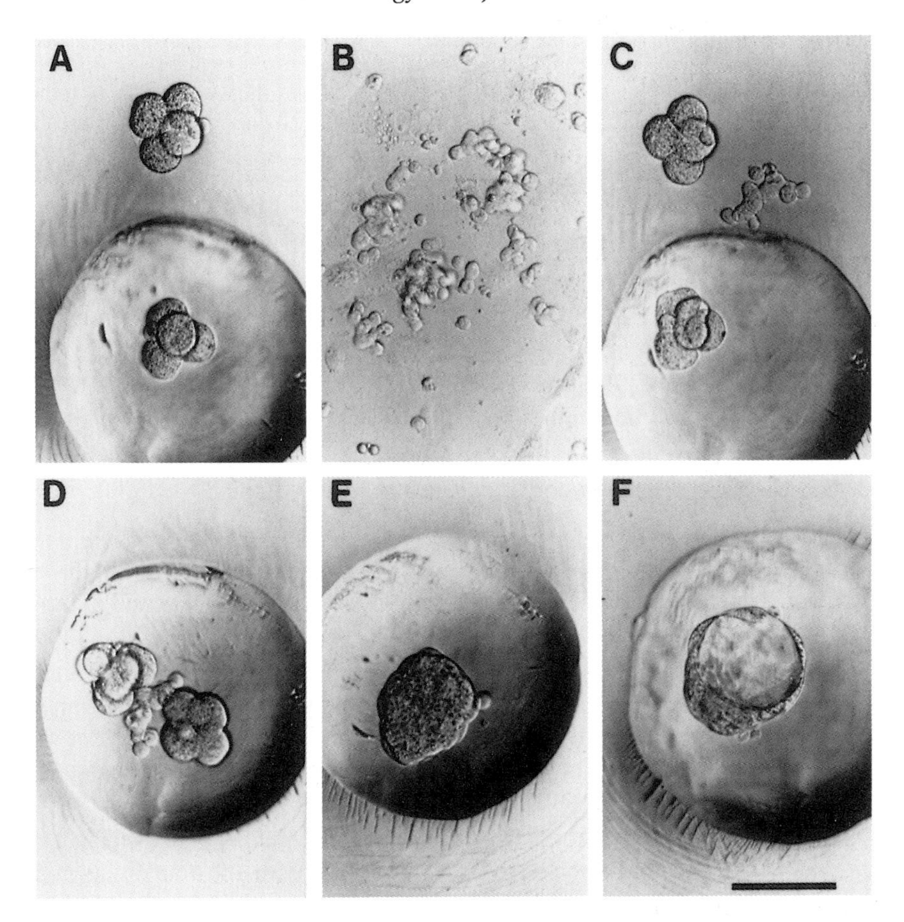

Figure 5. Steps of assembling 'sandwich' aggregates. Bar = 100 μm. (A) Two embryos placed in and beside the depression after removing their zona pellucida. (B) ES cells right after trypsin treatment. (C) The three components of a 'sandwich'. (D) 'Sandwich' assembled. (E, F) Aggregate forms compact morula then early blastocyst after overnight culture.

Protocol 10. Caesarean section

Materials

- fine scissors
- fine blunt-tipped forceps
- small pieces of soft facial tissue paper
- desk lamp

Protocol 10. *Continued*

Method

1. Kill the recipient by cervical dislocation on day 18.5 of pregnancy.
2. Open the body cavity by making a small ventral incision in the skin across the midline at the level of the bladder, then pull the skin up over the chest and cut the body wall along both sides starting from the midline at the bladder.
3. Hold the cervix with a pair of forceps, cut across the vagina below the cervix.
4. Carefully cut the uterine wall beside one fetus and let the fetus pop out.
5. Cut the yolk sac and amnion and quickly wipe the nose and mouth of the newborn.
6. Recover the other newborns by repeating steps 4 and 5.
7. Cut the umbilical cord after the placental circulation has stopped.
8. It helps the recovery of the newborns if you keep them in the palm of your hand under a desk lamp and frequently wipe their noses and mouths until they start breathing regularly and turn pink.
9. If required, take a blood sample, (e.g. for GPI assay to measure ES cell contribution) by cutting the tip of the tail and absorbing the little drop of blood at the wound with a small piece of tissue paper (4–5 mm^2). Put this tissue paper into an Eppendorf tube containing 20 μl of distilled water. Store at −20°C.
10. Remove the foster mothers from their cage and quickly take away half of their own pups. Mix the recovered newborns with the remaining pups and put the mothers back. It is recommended to have two foster mothers in the same cage.
11. After two to three hours, when the mothers have accepted the new babies, you can remove the remaining original pups. In this way no special marker is necessary to distinguish the original and the Caesarean-derived pups.

5. Characterization of mosaicism in diploid or tetraploid chimeras

In any experiments using chimeras one requires suitable markers to follow the fate of the components. This is especially important in ES cell–tetraploid chimeras, where the aim is to ensure that complete takeover of the fetus by ES cells has occurred. For many experiments with ES cell–diploid aggregation chimeras it will also be important to be able to follow the ES cell contribution in detail in different organs and tissues. For this reason it is advisable to use genetic markers other than simple coat colour differences in such experiments.

5.1 Marker systems

The most useful markers utilize natural or artificial genetic differences (genetic polymorphisms or transgenes) between the chimeric components. Here we describe three marker systems commonly used in connection with ES cell chimeras.

5.1.1 Glucose phosphate isomerase (GPI)

This is the most commonly used and most traditional chimeric marker, based on electrophoretic polymorphisms of GPI. The enzyme is a dimer of the products encoded by the two alleles of the gene *Gpi-1*. Therefore a mouse that is heterozygous for two different electrophoretic variants will show an electrophoretic pattern with three bands; the two homodimers and a heterodimer with intermediate mobility and double density. Three electrophoretically distinct alleles of the gene ($Gpi\text{-}1^a$, $Gpi\text{-}1^b$, and $Gpi\text{-}1^c$) have been described in laboratory mouse strains (see Chapter 4, *Table 1*). The products of these genes are designated as GPI-A, GPI-B, and GPI-C, respectively.

Most 129-derived ES cells, (e.g. D3, AB1) are GPI-A/A. If these cell lines are used, GPI-B/B (C57 × $CBAF_1$ or C57/BL) or GPI-A/B mice should be

Table 1. Preparation of major organs of a newborn for GPI assay.

Organs	Homogenate in sample buffer (μl)	Dilution with sample buffer
Brain	200	3 ×
Skin (0.5 cm^2)	200	2 ×
Adrenals	200	—
Kidneys	200	4 ×
Liver	200	8 ×
Stomach	200	2 ×
Pancreas	200	4 ×
Intestine	200	7 ×
Bladder	200	2 ×
Lungs	200	6 ×
Thyroid	200	2 ×
Spleen	200	2 ×
Testis	100	—
Muscle (deltoid)	200	2 ×
Heart	200	4 ×
Tongue	200	—
Thymus	200	—

used as a tetraploid partner. The occasional presence of weak GPI-B/B or GPI-A/B bands, respectively then indicates tetraploid contamination in ES cell-derived newborns. In chimeras with diploid embryos, where estimates of ES cell:host ratio are required the two components should be of different homozygous genotypes. With the GPI assay one can easily detect as low as 1–3% of contribution levels of the components. Different organs have different specific enzyme activity. *Table 1* shows suitable dilution factors or final volumes of the tissue homogenates of newborns to obtain equal activities for GPI cellulose acetate electrophoresis. Starch gel electrophoresis can also be used (Chapter 4, *Protocol 9*).

Protocol 11. GPI electrophoresis using cellulose acetate gel

Equipment (Helena Laboratories, 1530 Lindbergh Drive PO Box 752, Beaumont, Texas 77704)

- Super Z Applicator Kit (Cat. No. 4093), contains Super Z applicator (12 samples), Super Z aligning base, and Super Z well plate
- titan III cellulose acetate plates (Cat. No. 3024)
- disposable wicks (Cat. No. 5081)
- zip zone chamber (Cat. No. 1283)

Buffers

- sample buffer: 50 mM Tris–HCl pH 8.0 + 0.1% Triton X-100
- running buffer: Supre-heme buffer (Helena Laboratories, Cat. No. 5802), or 3 g Tris + 14.4 g glycine (Sigma Cat. No. G7126) made up to 1 litre with distilled water

Stock solutions for staining

	Stock concentration
(a) magnesium acetate 0.25 M	5.41 g/100 ml
(b) fructose-6-phosphate (Sigma Cat. No. F3627)	75 mg/ml
(c) tetrazolium (MTT) (Sigma Cat. No. M2128)	10 mg/ml
(d) phenazine methosulfate (PMS) (Sigma Cat. No. P9625) Store in light-proof container!	1.8 mg/ml
(e) nicotinamide adenine dinucleotide phosphate (NADP) (Sigma Cat. No. N0505)	10 mg/ml

(f)	glucose-6-phosphate dehydrogenase (G6PDH) (Sigma Cat. No. G8878)	50 U/ml
(g)	1% agarose (1 g/100 ml) in 0.2 M Tris–HCl pH 8.0	2.42 g/100 ml

Store stocks (a)–(f) in 200 μl aliquots at −20°C.

Preparation of samples

1. Freeze-thaw the samples at least twice in the appropriate volume of sample buffer. If the tissue volume is large enough, standard homogenization techniques can be used. For small samples simply press the tissue against the wall of the tube with the tip of a pipette.
2. Spin the samples and transfer 8 μl of supernatant to the wells of the Super Z well plate.

Electrophoresis

3. Prepare the chamber by filling the buffer compartments with Supre-heme buffer half-way and folding the wicks over the inner walls to electrically connect them to the buffer compartments.
4. Soak the cellulose acetate gel in Supre-heme buffer 15 min. prior to applying the samples by lowering the gel slowly into the buffer to avoid bubble formation on the surface.
5. Take the gel out and remove excess buffer by blotting between paper towels. Mark the plastic backing with the necessary information, e.g. gel number, anode.
6. Press down the applicator into the samples in the Super Z wells and then blot it on tissue paper. Return the applicator to the sample wells and hold it down in samples for a few seconds. Apply samples to gel by holding the applicator for a few seconds on the gel placed on the aligning base.
7. Transfer the gel (gel surface down) to the chamber and weight it down with a coin in the centre. Run at 300 V (from anodes to cathode) for 90 min.

Staining

8. While the gel is running boil 1% agarose in buffer and put it into a 55°C water-bath.
9. Just before the electrophoresis is complete, place 200 μl of each stock solution of stain components (a)–(e) into a test-tube (>12 ml) shielded from light with aluminium foil. Warm the mixture in the 55°C water-bath, then add 10 ml of 55°C agarose solution. Mix thoroughly.
10. Terminate gel run, blot the gel on tissue, and place the gel (gel side up) on a level surface.

Protocol 11. *Continued*

11. Add 200 μl of staining stock (f) (G6PDH) to the stain mixture from step **9**. Mix quickly and pour it evenly over the gel. To decrease background work in dim light, since the stain is light-sensitive.

12. Stain for 10–15 min. (depending on the intensity of the reaction) at 37°C, then fix the gel in 1:3 acetic acid:glycerol for 10 min.

13. Read the gel and make a photographic record.

5.1.2 DNA *in situ*

To date the most reliable ubiquitous cell marker in mice is a genomic DNA marker provided by a transgenic insertion of the mouse β-globin gene at extremely high copy number (~1000) (T-MβG-1 line of Lo(18)). DNA *in situ* hybridization to this large piece of repetitive DNA clearly marks the transgenic cells in tissue sections. In order to use this marker system either the host embryos or the ES cells should contain this transgene.

Protocol 12. Tissue fixing, embedding, and sectioning for DNA *in situ* hybridization

Materials

- poly-L-lysine (Sigma Cat. No. P1524)
- Tissue Prep 2 (Fisher Sci. Cat. No. T555)
- slides
- microtome

Method

1. Fix tissue in freshly prepared ethanol:acetic acid (3:1) at 4°C (half an hour for blood smears, a few hours for small embryos, overnight for large specimens)

2. Wash for 1 h in each of the following:
   - twice in ethanol at 4°C
   - ethanol:xylene (1:1) at room temperature
   - xylene at room temperature
   - xylene:Tissue Prep (1:1) at 60°C
     Mix well with gentle inversion
   - Tissue Prep at 60°C

3. Transfer tissue into new Tissue Prep (60°C) for 1 h then make blocks.

Poly-L-lysine coating of slides

4. Make up 200 ml 50 μg/ml poly-L-lysine solution in a plastic beaker, stir to dissolve.

5. Immerse slides for 5 min, then air-dry on a rack. Use the slides the next day. (Maximum shelf-life is one week.)

Sectioning and dewaxing

6. Section tissues at ~7 μm.

7. Dewax slides in the following washes for 10 min. each: xylene, xylene, xylene:ethanol (1:1), ethanol, ethanol).

8. Allow to air-dry and store until ready to use.

Protocol 13. DNA *in situ* hybridization with biotinylated probes

Solutions

- streptavidin–horseradish peroxidase: Boehringer Mannheim No. 1089 153 (500 U/ml); store at 4°C
- biotin-labelled probe to sequence of interest; labelled by standard procedures
- dextran sulfate/SSPE mix (1 g dextran sulfate/2 ml of 20 × SSPE diluted to 6 ml, aliquoted, and stored at 4°C)
- 30% H_2O_2 (Sigma H1009) stored at −20°C
- diaminobenzidine, DAB (Sigma D5637) stock. Dissolve 10 g DAB in 200 ml 50 mM Tris pH 7.2 or water. Store in 2 ml aliquots at −20°C. Prior to use, thaw one tube, dilute up to 200 ml in 50 mM Tris pH 7.2, add 67 μl 30% H_2O_2, and store at −20°C.

Note: DAB is a carcinogen. Wear gloves when handling, rinse all glassware with bleach, then water.

- buffers (see (19) for recipes), PBS, 20 × SSC, SSPE

Methods

1. To remove endogenous peroxidases, immerse slides for 30 min. in methanol + 30% H_2O_2, then rinse several times in PBS.

2. Denature genomic DNA by immersing the slides for 2 min. in 70% formamide in 2 × SSC at 70°C, then plunge into cold 70% ethanol (stored at −20°C) for 2 min., and dehydrate through 75%, 80%, 90%, 95% ethanol for at least 2 min. each, then air-dry.

Protocol 13. *Continued*

3. Meanwhile prepare hybridization mix (allow 50 μl per slide). The volumes should be multiplied by the number of slides.
   (a) Remove an aliquot of 20 μl biotinylated probe (1.6 μg/ml) and boil it for at least 5 min., then place immediately on ice.
   (b) Add 30 μl of dextran sulfate/SSPE mix to cooled DNA. Pipette up and down to mix, keep on ice.
4. Add hybridization mix to denatured slides, remove any bubbles over sections, and cover with coverslip (22 × 60 mm covers entire slide area).
5. Place slides in a moist chamber overnight at 60°C.
6. Wash slides in the following:
   - 2 × SSC + 0.1% Triton X-100 for 5 min. at room temperature
   - 1 × SSC + 0.1% Triton X-100 for 5 min. at room temperature
   - 1 × SSC + 0.1% Triton X-100 for 10 min. at 50°C
   - 10 mM Tris in 1 × PBS + 1% BSA (TBS + BSA). (Check pH of final solution. It should be between 7.2 and 7.6. Adjust with HCl if necessary.)
7. Dilute stock streptavidin–horseradish peroxidase 1:1000 in TBS + BSA, then incubate slides with it for 2 h at 35°C. (Use 100 μl/slide.)
8. Wash twice in TBS + BSA for 5 min. each.
9. Thaw one tube of DAB stock, dilute to 200 ml in TBS + BSA, and add 67 μl 30% H_2O_2 (stored at −20°C).
10. Incubate slides in the freshly prepared DAB + H_2O_2 for 10 to 60 min. Cover container with foil, place away from direct light during incubation.
11. Rinse slides briefly in TBS.
12. Stain the sections (usually with haematoxylin-eosin) lightly.
13. Cover slides with Permount and examine for hybridization signal.

Note: DAB signal stains reddish-brown. Addition of 1% nickel chloride or 1% cobalt chloride slightly enhances signal, and changes the signal colour to dark blue.

5.1.3 *LacZ* gene expression

The bacterial β-galactosidase enzyme encoded by the *lacZ* gene provides an easily detectable marker for transgenic expression using simple histochemical staining procedures. This gene, under the control of suitable promoters, can be used either to mark the ES cell or the embryo component of chimeras. The staining procedures for different embryonic stages are found in Chapter 6, Section 4.3.

5.2 Purity of ES cell-derived embryos from tetraploid aggregates

Analysis of embryos derived from ES cell–tetraploid aggregates, using the genetic markers above, has revealed that the primitive ectoderm-derived lineages, like yolk sac mesoderm, amnion, and the embryo proper were completely ES cell-derived (2). The ES cell lines that we have used appear to be limited in their ability to contribute to extra-embryonic membranes. Therefore, when they were aggregated with tetraploid embryos, the tetraploid component provided the trophoblastic lineages which formed the major component of the placenta and the primitive endoderm derivatives that form the endoderm layer of the yolk sacs. Occasionally there was some tetraploid contamination in the fetus itself. This is not very surprising, since it was shown earlier that tetraploid cells could be rescued by the diploid cells in the embryonic lineages of aggregation chimeras at a low frequency and at a low level of contribution (20). In our hands, about one out of ten ES cell–tetraploid aggregates that reached term contained a trace amount (less than 5%) of tetraploid cells as judged by GPI. In experiments where the purity of ES cell-derived fetuses is crucial you should use markers, (i.e. GPI) to check tetraploid contamination.

5.3 Contribution of ES cells to different organs of diploid chimeras

An increasing number of genetic manipulation techniques utilize the ability of ES cells to differentiate into somatic tissues. The validity of these approaches depends on the ES cell-derived cells showing the same developmental and functional behaviour as embryonic cells. We have shown that ES cells and ICM cells of the same genotype contribute to different organs of the chimeras in a similar manner (21, 22). ES cell-derived cells were not specifically excluded from any of the different somatic lineages in chimeras, although there were significant differences in the contributions of ES cells among the organs. The organ-specificity of this effect depended on the genotype combination used in the ES cell–embryo aggregates, and was the same as that with ICM–embryo aggregates of the same genotypes. Thus, genotype effects can favour the contribution of ES cells to specific lineages. This phenomenon could explain why certain combinations of ES cells and host genotypes, (e.g. 129-derived ES cells and C57BL/6J host) are better at producing a germline contribution of ES cells than others.

6. Pre-natal development and perinatal mortality of ES cell–diploid and ES cell–tetraploid embryos

Table 2 shows the pooled data of pre-natal development of aggregates from three cell lines (A3(Merentes-Diaz, Gocza, and Nagy, unpublished), D3(8), and AB-1(9)) to provide an idea of the success rates for those experiments. On average, 41% of the implanted ES cell–diploid aggregates reached term, with 57% containing ES cell contribution. ES cell–tetraploid aggregates implanted at a similar rate, but they showed an increased rate of early post-implantation death (resorption in the table). This probably reflects aggregates that failed to internalize ES cells, since pure tetraploid embryos mostly die shortly after post-implantation (3, 13). Only 18% of implanted ES cell–tetraploid aggregates reached term, but almost all were completely ES cell-derived. However, birth was another major time of lethality. Except early passages of R1(7), all the other ES cell line-derived newborns produced died at term. In terms of birth weight, gross morphology, and anatomy they were apparently normal. However, following Caesarean section or birth they failed to oxygenize their lungs and died.

Perinatal mortality was found not only in connection with ES cell–tetraploid chimeras (completely ES cell-derived newborns) but also with ES cell–diploid chimeras having high ES cell contributions (2, 21). In *Figure 6* the perinatal fate of the 42 ES cell–diploid chimeras listed in *Table 2* is shown as a function of average ES cell contribution. We have not achieved post-natal survival when fetuses contained greater than 80% ES contribution with cell lines other than R1.

Beside D3, AB-1, and A3, we have tested the pluripotentiality of 11 other lines established in our laboratories. These lines also covered three different genotypes; 129/Sv, C57/BL, and C57 × $129F_1$. Five cell lines did not allow development to term when aggregated with tetraploid embryos. Karyotype analysis on these lines showed that they had only between 40% to 60% normal mitotic spreads (Merentes-Diaz and Nagy, unpublished). Six of the 11 lines were able to support completely ES cell-derived development up to term to a certain extent (2–15% of aggregates transferred). We found a normal number of 40 chromosomes in 70% to 80% of the mitotic spreads in these cell lines and no difference in karyotype was seen. R1 was the only one which gave rise to survival of completely R1-derived newborns and resulted in normal, fertile males (7). Even at early passage of R1, however, there was a sign of decreased developmental potential. Only one third of newborns recovered and one fourth reached adulthood (*Table 3*). At late passage number R1 behaved like the other cell lines which supported development to term but the ES cell-derived newborns were not able to recover from Caesarean section (*Table 3*).

7. Factors affecting pluripotentiality of ES cells

Twelve years after ES cells were first described (23, 24) their nature is still obscure. They are permanent cell lines and thus resemble tumour cells. ES cells do not, however, form tumours at high frequency if they are introduced back into a mouse in a chimera. Although a proportion of newly established

Table 2. Pre-natal development of diploid–ES cell and tetraploid–ES cell aggregates

Type	Number of aggregates transferred	Number of implantations (% of transferred)	Number of resorptions (% of implant.)	Number of midgestation dead (% of implant.)	Number of newborns (% of implant.)	Number of recovered (% of implant.)
ES cell–diploid	217	181 (83)	74 (41)	33 (18)	74 (41)	42 (57)
ES cell–tetraploid	307	191 (62)	119 (62)	38 (20)	34 (18)	34 (100)

Table 3. Development of ES cell–tetraploid aggregates made from early and late passages of R1 cell line.

Passage number of R1	Number of tetraploid aggregates transferred	Number of resorptions (% of transferred)	Number of midgestation dead (% of transferred)	Number of newborns (% of implant.)	Number of recovered (% of implant.)	Number reached adulthood (% of transferred)
5–14 passages	130	45 (35)	22 (17)	20 (15)	9 (7)	5 (4)
15–24 passages	162	51 (31)	46 (28)	12 (7.4)	1 (0.6)	0 (0)

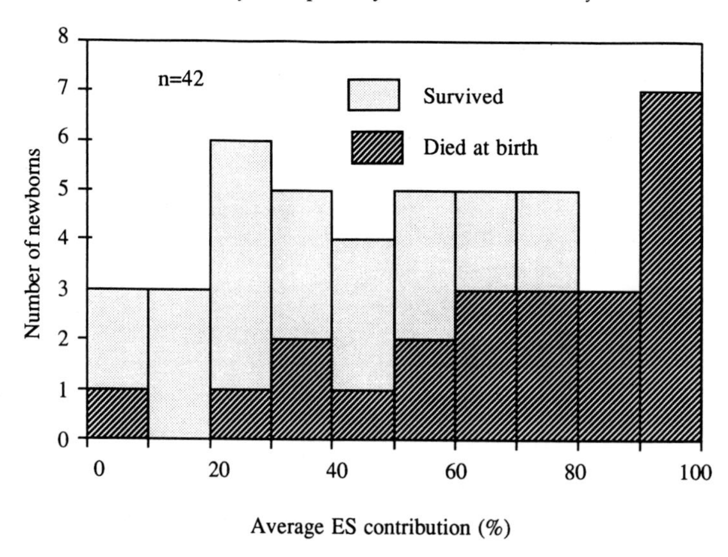

Figure 6. Relationship between the extent of ES cell contribution and survival of newborns. (Data were pooled from experiments with D3, AB-1, and A3 shown in *Table 2*.)

ES cell lines have abnormal karyotypes, the majority maintain the normal chromosome number quite well. A normal karyotype is clearly important for pluripotentiality but is not the only factor. The culture conditions, as well as the genetic and epigenetic defects that accumulate during the course of *in vitro* culture also likely contribute to a decrease in developmental potentiality.

7.1 Culture conditions

The proper culture requirements of ES cells are complex. The relatively high concentration of FCS used is presumably a rich source of undefined essential factors. It is always wise to test several lots of FCS for plating efficiency and then buy the best lot in a large quantity (see Chapter 2, Section 2.4). In general, ES cells are very sensitive to culture conditions; small changes in their environment may induce their differentiation and impair their pluripotentiality. It is always useful to test the pluripotentiality of ES cell lines in ES cell–tetraploid chimeras immediately after you obtain or establish them. Then keep the cells in culture for another 10–15 passages and test them again. The comparison of early and late passage number cells can determine whether your culture conditions are sufficient to maintain the original pluripotentiality. You may expect, however, some decrease in pluripotentiality even under optimal culture conditions, since the cells may accumulate deleterious genetic, (i.e. chromosome loss) or epigenetic changes during proliferation.

7.2 Accumulating genetic and epigenetic defects

Chromosomal anomalies; chromosome loss, translocation, deletion, and mutations are certainly major sources of deleterious genetic changes in embryonic stem cells. There is not enough data available to estimate the average amount of damage due to these events that might occur during a defined period of culture. One might expect that different chromosomes are not equally sensitive to changes and that certain types of changes will be subject to negative or positive selection in culture. The picture is presumably quite complex. Genetic changes of these sorts are irreversible and so they will accumulate during *in vitro* culture. Such genome damage may significantly affect pluripotentiality of ES cells and their ability to undergo germline transmission.

Epigenetic changes have to be taken into account as well. From the example of genomic imprinting one might expect that similar mechanisms operate not only on the parentally differentially imprinted genes but on the other parts of the genome affecting the default expression levels of both alleles of different genes. Changing these epigenetic modifications of the genome may negatively influence the general pluripotentiality of ES cells. However, such defects can be reversed when the ES cells go through the germline.

7.3 Subcloning ES cells

Since ES cells may accumulate genetic and epigenetic changes during *in vitro* culture there is a risk that some subclones may have a severely restricted potential and, as a consequence, will not support fetal development or go through the germline. This may be because the original single cell that was cloned contained a genetic abnormality or because a defect occurred in one cell in the subcloned population soon after cloning. The proportion of sublines whose potential resembles that of the original parental line is a function of the particular cell line, the number of passages of the line, and the particular culture conditions used. For many experiments it is critical that the proportion of fully pluripotent subclones be high. When conditions are optimal this can be achieved. For example, we have shown that three subclones of D3 obtained after retrovirus infection, G418 selection, and 46 days in culture, all retained the ability to produce completely ES cell-derived 15.5 day embryos from ES cell–tetraploid aggregates at a frequency comparable to parental D3 cells (Chambers, Kang, Rossant, Hozumi, and Nagy, unpublished). Furthermore one of five sublines generated from passage 12 cells of R1 has the original capability of R1 since it can give rise to completely R1-derived viable mice from ES cell–tetraploid aggregates (7).

8. Applications for completely ES cell-derived embryos and fetuses

Numerous powerful genetic manipulations can be applied to mouse ES cells (review in 25). ES cells provide an alternative way to zygote injection for introducing foreign DNA, (e.g. dominant-acting expression constructs) into the mouse genome. They can also be used to apply novel strategies to identify and mutate genes having important and specific roles in development and differentiation (for example see Chapter 6). The most important technology applied to ES cells to date is the targeted mutagenesis of genes of interest, which provides a very effective way of determining the role of individual genes and eventually reconstructing the genetic interactions that go on during complex developmental processes. This latter approach requires the transmission of targeted ES cells through the germline in order to analyse the resulting phenotype of mice homozygous for the mutation. The current low frequency to achieve perinatal survival of completely ES cell-derived newborns means that this approach can not yet be applied readily to increasing the efficiency of germline transmission. Most of the cell lines, however, are able to support the perinatal development. The availability of completely ES cell-derived embryos and fetuses provides a large variety of applications for studying the effects of dominant-acting manipulations on embryonic development.

8.1 Direct analysis of dominant-negative, gain of function, and homozygous double-knockout mutations

Germline transmission of genetically manipulated ES cells is usually not an easy task. Not only because of the length of time that it requires, but also because manipulations that act dominantly can interfere with normal development, germline transmission, or production of live mature chimeras. Studying completely ES cell-derived development could not only accelerate but also extend the analysis of dominantly-acting genetic changes, such as over or ectopic expression of certain genes, and interference with the endogenous gene function by antisense or dominant-negative constructs. ES cell-derived fetuses can also be used to study phenotypes of homozygous mutations produced by targeting both alleles of a gene in ES cells.

Enhancer and gene trap strategies are the other rapidly developing areas of ES cell genetic manipulations (26 and Chapter 6). In both, the transgene serves as a reporter of endogenous gene activity. As a first screen for interesting patterns of 'trapped' gene activity, ES cell injection chimeras are usually analysed. It is obvious that the higher the ES cell contribution, the less likely it is that host cells will interfere with visualization of expression pattern. Therefore the use of ES cell–tetraploid chimeras is a logical choice in this screening approach.

8.2 Source of tissues and tissue-progenitors

As well as studying ES cell-derived fetal development itself, the derivation of completely ES cell-derived fetuses allows access to a rich source of differentiated tissues and tissue-progenitors, which can carry the genetic manipulation of interest. Dissection of such tissues from ES cell-derived embryos or newborns allows characterization of their phenotype either in culture or *in vivo* by transplanting them back to a suitable recipient. As an example of such an application we have shown (27), that fetal liver cells of 15.5 day completely ES cell-derived embryos were able to reconstitute the haemopoietic system of lethally irradiated adult recipients. Two to three months after grafting, all the haemopoietic lineages of these recipients were completely ES cell-derived. This procedure allows introduction of a variety of genetic alterations exclusively into the adult haemopoietic system.

Protocol 14. Reconstitution of the mouse haemopoietic system by ES cell-derived fetal liver cells

Materials

- fine surgical instruments for dissecting embryos: scissors, forceps (1.5 mm tip)
- DMEM + FCS medium
- 1 ml syringes with 23 and 27 gauge needle
- sterile test-tubes

Method

1. Dissect ES cell–tetraploid embryos on day 15.5.
2. Remove fetal livers and put them individually into a test-tube containing 1 ml of DMEM medium.
3. Disaggregate the liver by drawing the clumps of cells several times through a 23 G needle.
4. You should check the purity of the ES cell-derived liver by using a GPI marker to distinguish the tetraploid and ES cell components. Quickly run a gel (see *Protocol 11*) with 10 μl of the cell homogenate (freeze-thaw this 10 μl cell suspension before applying to the gel).
5. While the gel is running, irradiate the recipients. The proper dose varies from strain to strain, (e.g. 129/Sv, 9.5 Gy; C57BL/6J and C57BL/6J × 129/Sv F_1, 10.5 Gy).
6. Count cells and prepare 10^7 cells/ml DMEM cell suspension.
7. Inject 5×10^6 cells IV into the tail vein of the recipients, using a 27 G needle.

ES cell-derived haemopoiesis may not be the only system accessible to this approach of rescue by transplantation into adult mice. Similar methods may be applicable to transplantation of skin, neural tissues, and in the case of female ES lines, ovaries.

9. Prospects and limitations

The ability to take cells from tissue culture and reconstitute an entire fetus or an animal is a remarkable testament to the stability and pluripotency of ES cells. We have tended to emphasize the difficulties in achieving viable fully ES cell-derived or extensively ES cell-derived live-born mice after embryo aggregation, in order to prepare the novice for some disappointments. However, even with existing established ES cell lines, it is possible to use the aggregation approach outlined here to produce ES cell-derived fetuses, which can be used for a variety of studies on the effects of genetic manipulations on somatic lineages. It seems likely that, in the future, as we learn more about the optimal conditions for deriving and maintaining ES cells, that fully ES cell-derived animals will be more readily and reproducibly achieved. Whether it will be possible to maintain ES cell potency through extensive passages in culture, will depend on determining the nature of the changes that lead to loss of full potency, and then learning how to prevent them.

Acknowledgements

We wish to thank our colleagues; Wanda Abramow, Alan Berstein, Cynthia Chambers, Lesley Forrester, Elen Gocza, Nobumichi Hozumi, Eszter Ivanyi, Joon-Soo Kang, Lajos Laszlo, Merja Markkula, Elizabeth Merentes-Diaz, Reka Nagy, Valerie R. Prideaux, and John Roder for their contributions. This work was supported by grants from the Medical Research Council of Canada, the National Institute of Health (OTKA#2308) and OMFB (Hungary). J.R. is a Terry Fox Cancer Research Scientist of the NCIC and an International Scholar of the Howard Hughes Medical Institute.

References

1. Beddington, R. S. and Robertson, E. J. (1989). *Development,* **105,** 733.
2. Nagy, A., Gocza, E., Merentes-Diaz, E., Prideaux, V. R., Ivanyi, E., Markkula, M., and Rossant, J. (1990). *Development,* **110,** 815.
3. Kaufman, M. H. and Webb, S. (1990). *Development,* **110,** 1121.
4. Tarkowski, A. K., Witkowska, A., and Opas, J. (1977). *J. Embryol. Exp. Morph.,* **41,** 47.
5. Stewart, C. L. (1982). *J. Embryol. Exp. Morph.,* **67,** 167.
6. Wagner, E. F., Keller, G., Gilboa, E., Ruther, U., and Stewart, C. (1985). *EMBO J.,* **3**(4), 663.
7. Nagy, A., Rossant, J., Nagy, R., Abramow-Newerly, W., and Roder, J. (1993). *Proc. Natl Acad. Sci. U.S.A.,* in press.
8. Doetschman, T., Eistetter, H., Katz, M., Schmidt, W., and Kemler, R. (1985). *J. Embryol. Exp. Morph.,* **87,** 27.
9. McMahon, A. P. and Bradley, A. (1990). *Cell,* **62,** 1073.
10. Quinn, P., Barros, C., and Whittingham, D. G. (1982). *J. Reprod. Fert.,* **66,** 161.

11. Whittingham, D. G. (1971). *J. Reprod. Fertil. (suppl.),* **14,** 7.
12. Edwards, R. G. (1958). *J. Exp. Zool.,* **137,** 309.
13. Snow, M. H. (1973). *Nature,* **244,** 513.
14. O'Neill, G. T., Speirs, S., and Kaufman, M. H. (1990). *Cytogenet. Cell Genet.,* **53,** 191.
15. Eglitis, M. A. (1980). *J. Exp. Zool.,* **213,** 309.
16. Kubiak, J. Z. and Tarkowski, A. K. (1985). *Exp. Cell Res.,* **157,** 561.
17. Hogan, B., Constantini, F., and Lacy, E. (1986). *Manipulating the mouse embryo.* Cold Spring Harbor Press, Cold Spring Harbor, NY.
18. Lo, C. W. (1986). *J. Cell Sci.,* **81,** 143.
19. Maniatis, T., Fritsch, E. F., and Sambrook, J. (ed.) (1982). *Molecular cloning. a laboratory manual.* Cold Spring Harbor Press, Cold Spring Harbor, NY.
20. Lu, T.-Y. and Markert, C. L. (1980). *Proc. Natl Acad. Sci. U.S.A.,* **77,** 6012.
21. Rossant, J., Merentes-Diaz, E., Gocza, E., Ivanyi, E., and Nagy, A. (1993). In *Serono symposium on preimplantation embryo development* (ed. B. Barister), Springer-Verlag, NY.
22. Merentes-Diaz, E., Gocza, E., Ivanyi, E., Gerlai, R., Rossant, J., and Nagy, A. (1993). Submitted.
23. Evans, M. J. and Kaufman, M. (1981). *Nature,* **292,** 154.
24. Martin, G. R. (1981). *Proc. Natl Acad. Sci. U.S.A.,* **78,** 7634.
25. Rossant, J. (1990). *Neuron,* **2,** 323.
26. Skarnes, W. C. (1990). *Biotechnology,* **8,** 827.
27. Forrester, L. M., Bernstein, A., Rossant, J., and Nagy, A. (1991). *Proc. Natl Acad. Sci. U.S.A.,* **88,** 7514.

6

Gene and enhancer trap screens in ES cell chimeras

ACHIM GOSSLER and JOCHEN ZACHGO

1. Introduction

LacZ reporter vectors in combination with mouse embryonic stem (ES) cells provide valuable new approaches for detecting patterns of gene expression during mouse embryogenesis and for identifying, isolating, and disrupting (mutating) endogenous genes. Reporter constructs can also be introduced into other cell types with a more restricted developmental potential, such as haemopoietic stem cells, to specifically identify genes expressed or repressed during particular cellular differentiation pathways.

In this chapter we focus on the use of 'trap' vectors in ES cells. In addition to the practical aspects of this approach, the properties of various types of trap vectors are compared. The implications of these properties for the use of each vector in various experimental conditions is discussed as well as some possible improvements and alterations.

1.1 General properties of enhancer, gene, and promoter trap vectors

Three general types of 'trap' vectors have been devised and we refer to them as enhancer trap (ET), gene trap (GT), and promoter trap (PT) vectors. Depending on the authors, gene traps are also called promoter traps (see footnote in *Table 2*). The main difference between the vectors is with respect to the modes of activation of the reporter gene. Throughout this chapter we use a terminology solely based on the presence or absence of various sequence elements in the constructs that confer the different requirements for activation on the reporter gene. In most cases the reporter gene is the *E. coli lacZ* gene whose expression can be easily detected by the enzymatic activity of its product β-galactosidase. We refer to enhancer traps as constructs which carry a reporter gene with a start of translation (ATG codon) under the control of a minimal promoter. Gene traps are vectors that carry the reporter gene (with or without an ATG codon) 3′ to a splice acceptor site, whereas promoter traps contain only a reporter gene (with or without an ATG) with neither a

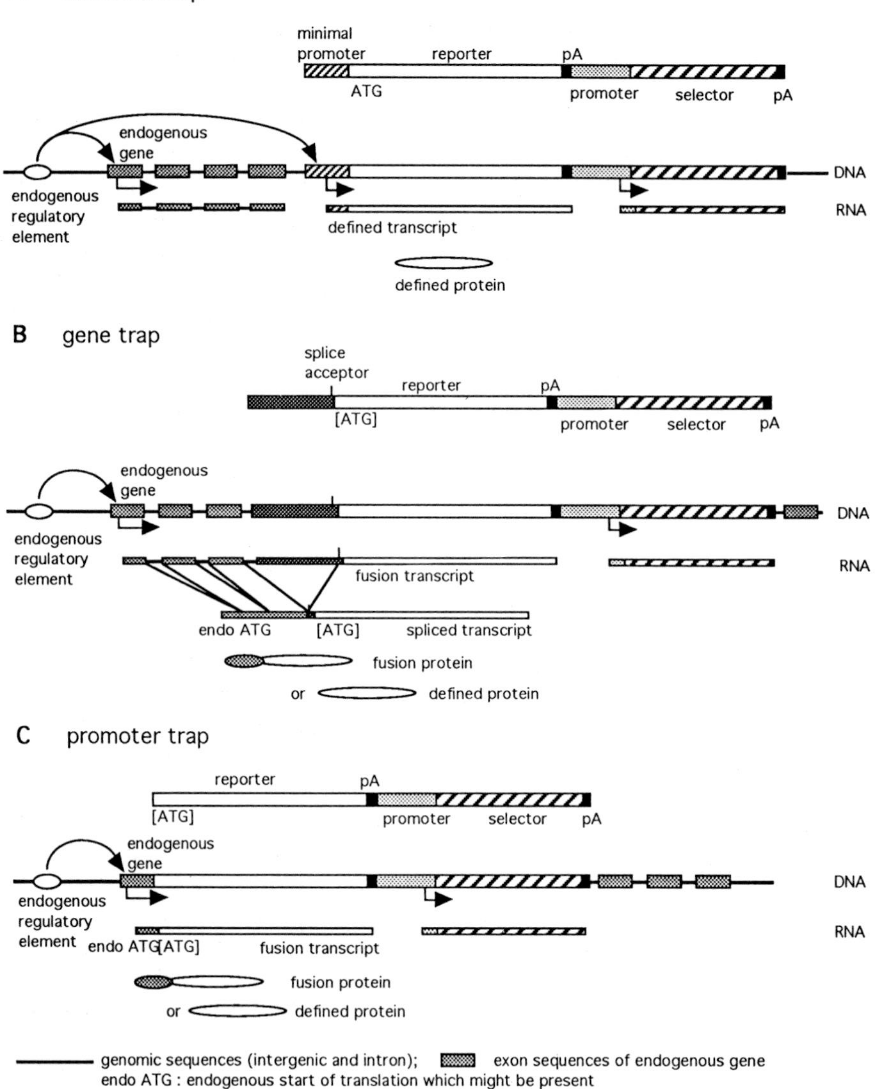

Figure 1. Principle structure and mode of activation of different reporter gene constructs. For simplicity only one of several possible integrations that could give rise to functional β-galactosidase has been drawn for each vector-type. (A) Enhancer trap: a minimal promoter is placed in front of the *lacZ* gene. Integrations within or outside the endogenous gene in both orientations should result in *lacZ* gene expression. Transcription starting from the minimal promoter results in defined transcripts and a defined protein. (B) Gene trap: a splice acceptor is placed in front of the *lacZ* gene. Integration into an intron in the correct orientation results in a primary *lacZ* fusion transcript which is further processed. If no ATG is provided with the *lacZ* gene then the spliced transcript can only give rise to functional

promoter or splice acceptor sequence (see *Figure 1, A–C*). Enhancer and gene trap vectors have been used primarily thus far in ES cells to screen for genes expressed during mouse embryogenesis and emphasis will be put on these vector-types.

The activation of transcription of enhancer trap vectors is based on position effects which are frequently exerted on the expression of genes introduced into the genome of eukaryotes. A typical enhancer trap construct contains the *lacZ* gene under the control of a weak (minimal) promoter. Upon integration into the genome, the promoter is frequently activated by endogenous *cis*-acting elements close to or at the site of integration (see *Figure 1A*), thus leading to expression of the *lacZ* gene. Enhancer trap vectors have been used to visualize transcriptional activation patterns and to identify and isolate genes expressed in spatially and temporally regulated patterns during development in *Drosophila*. Using P element-mediated transposition numerous *Drosophila* strains carrying enhancer trap integrations in their genome have been generated and have been analysed for β-galactosidase staining patterns (for example 1, 2). A number of *Drosophila* genes have been isolated from enhancer trap integration sites and in many cases the β-galactosidase staining patterns reflect the expression pattern of the endogenous gene (3). In the mouse, *lacZ* enhancer trap vector expression was first analysed in transgenic mice produced by DNA microinjection (4, 5). However, isolation of the corresponding endogenous genes in such transgenic mice can be hampered in many cases by multiple copy integrations and/or structural rearrangements that occur at the site of integration following DNA microinjection. Thus far, from none of the transgenic lines showing *lacZ* staining patterns (six lines in the reports cited above) has the isolation of a corresponding endogenous gene been reported. These problems can to some extent be overcome by the use of embryonic stem (ES) cells, since electroporation can allow for single copies of the vector to be integrated and in our studies of four integration sites, no major rearrangements of the endogenous locus were observed. In addition, ES cells allow for an increase in the number of integration sites that can be analysed for transcriptional activation patterns during early mouse post-implantation development. The strategy employing mouse embryonic stem

---

β-galactosidase when a start of translation signal is present in an upstream exon and the reading frame is maintained during splicing. If an ATG is included at the 5′ end of the *lacZ* gene, this can be used as a start of translation if translation is not initiated at an endogenous ATG further upstream. In cases in which an upstream endogenous ATG is used then again the reading frame has to be maintained. Integration into an exon might also give rise to functional β-galactosidase if an ATG is present for the start of translation. (C) Promoter trap: no promoter or splice acceptor sequences are present 5′ to the *lacZ* gene. Integration into an exon in the correct orientation downstream of a start of transcription gives rise to a fusion transcript. Depending on the presence or absence of an endogenous ATG and an ATG provided within the construct, functional β-galactosidase may be generated as a fusion protein or defined protein.

(ES) cells involves introduction of a *lacZ* enhancer trap vector, that includes a selectable marker gene, into ES cells by electroporation and isolation of drug resistant cell clones which have taken up DNA. Clonal cell lines are then used to produce chimeric embryos which after subsequent development in foster females are stained for β-galactosidase activity (see *Figure 2*) (6). In this way, hundreds of cell lines representing independent vector integrations can be analysed efficiently.

In addition to enhancer trap constructs, gene trap-type vectors have been used in ES cells (6–8). As mentioned above, gene trap vectors do not contain a promoter in front of the *lacZ* gene but a splice acceptor site. With a basic gene trap vector functional β-galactosidase can only be expected when integra-

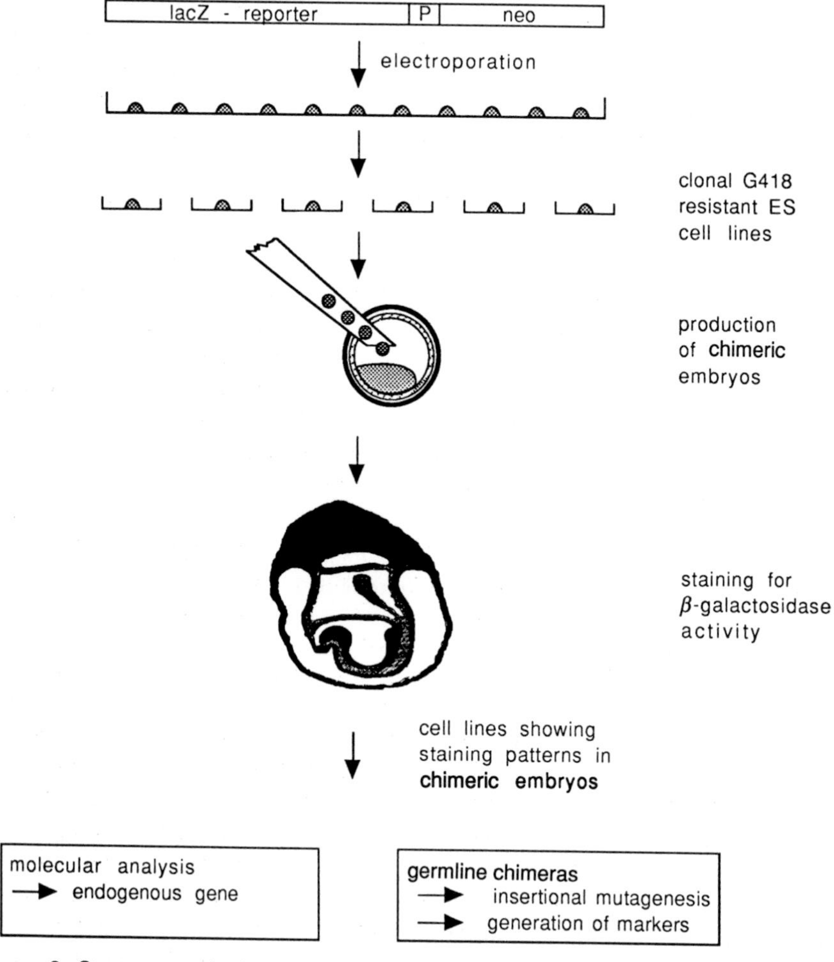

Figure 2. Strategy employing embryonic stem cells to identify transcriptional activation patterns during embryogenesis.

tions occur in introns allowing upstream endogenous exons to be spliced to *lacZ*. This produces fusion transcripts, some of which maintain the reading frame between the upstream exon and *lacZ* producing functional *lacZ* protein fusion products. When an ATG is provided for the *lacZ* gene, then translation can be initiated at this ATG if a translation initiation site is not present in the upstream (endogenous) sequences (see *Figure 1B*).

Promoter trap vectors have been used to detect patterns of gene expression in *Caenorhabditis elegans* (9), to identify and isolate promoters in fibroblasts (10–12), and have also recently been applied to identify and mutate genes expressed in ES cells (13). They require integration into exons. As in gene traps, fusion transcripts between cellular and vector sequences are generated but no splicing is required for expression of functional β-galactosidase. The part of cellular transcribed sequences fused to *lacZ* can be extremely short if the integration occurs in close proximity to the start of transcription (see *Figure 1C*). In principle, a gene trap construct could behave like a promoter trap if the integration occurs into the first exon. However, in the gene trap integrations that have been analysed at the molecular level, the *lacZ* gene integrated into introns (8).

The different 'targets' in the genome for the different trap vector-types, and thus the different modes of activation, have implications for the use of these constructs and for the experimental design. The three vector-types offer various advantages and disadvantages with respect to frequency of productive integrations, access to the endogenous gene, and potential mutagenicity (see *Table 1* for a comparison between ET, GT, and PT vectors).

The relative efficiency of different constructs can be determined by comparing the number of ES cell clones displaying β-galactosidase activity ('blue lines') with the number of ES cell clones carrying the reporter construct and thus showing resistance to the selectable drug ('neo^R lines'). The requirements to obtain a functional *lacZ* gene product from gene and promoter trap integrations are more stringent than for enhancer trap integrations. Thus, the frequency of integration events leading to detectable β-galactosidase activity is higher with enhancer traps. Analysis of *lacZ* expression in undifferentiated ES cell clones has shown, that 20–25% of neo^R ET lines express detectable levels of β-galactosidase activity whereas only 1–5% of GT lines express *lacZ*. This suggests, that about 5–20 times fewer GT than ET integrations into the genome are 'productive' and can give rise to functional β-galactosidase. According to these numbers, 5–20 times as many GT as ET integrations have to be analysed to detect a staining pattern during early embryogenesis. Therefore, thus far only GT ES cell lines expressing functional β-galactosidase have been analysed in chimeric embryos. Due to the constraint of integration into an exon in the correct orientation and often reading frame, PT vectors are expected to be activated at a lower frequency than GT vectors. In one set of experiments with a PT retroviral vector in NIH 3T3 cells, the *lacZ* gene was found to be activated in 0.4% of the lines (12). In ES cells, only PT vectors

Table 1. Comparison of ET, GT, and PT properties

	Enhancer trap	**Gene trap**	**Promoter trap**
Requirements for expression of functional β-galactosidase	Integration close to *cis*-acting elements able to activate the minimal promoter	Integration most likely in an intron with orientation and reading frame dependence	Integration in an exon with orientation and reading frame dependence
Transcript	Defined start site in promoter	Fusion transcript between endogenous gene and *lacZ*	Fusion transcript between endogenous gene and *lacZ*
Physical linkage to endogenous gene exons	Genomic DNA adjacent to integration site	Fusion mRNA or genomic DNA adjacent to integration site	Fusion mRNA or genomic DNA adjacent to integration site
Mutagenesis	Possible	Very likely	Very likely
Requirement for target gene expression in ES cells	No	Yes	Yes

carrying marker genes that are directly selectable have been used at present (13).

The higher frequency of productive integrations with enhancer trap vectors means that ET lines either expressing ('blue lines') or not expressing ('white lines') detectable levels of *lacZ* in undifferentiated stem cells can be analysed in chimeric embryos (see *Table 2*). Various modifications of gene trap vectors can be envisaged to increase the overall frequency of integration events leading to expression of functional β-galactosidase (see below). This could then allow analysis of 'white' GT ES cell lines. However, at present, considering the frequency of integrations showing staining patterns in the embryo without pre-selection for genes active in ES cells, enhancer trap constructs are more efficient than gene trap and promoter trap vectors. On the other hand, the mechanism of GT and PT activation provides these vectors with two advantages which for most purposes outweigh the disadvantage of the lower efficiency of gene traps to detect patterns of gene expression.

First, by definition, GT and PT integrations that express detectable levels of β-galactosidase (or other markers) must have occurred within a gene, and endogenous transcribed sequences must be physically linked to the *lacZ* (marker gene) mRNA in the fusion transcript, providing direct access to the endogenous gene on the cDNA level. In contrast, *lacZ* expression from enhancer trap constructs integrated into the genome starts at the minimal promoter of the vector. The regulatory element(s) exerting its (their) influence on the promoter could be located on either side and even at some distance to the integration site. The link between the observed pattern and the corresponding endogenous gene is the integration site. Thus, the genomic sequences surrounding the ET vector have to be cloned and have to be analysed to identify transcribed sequences.

Secondly, GT and PT integrations that lead to *lacZ* (marker gene) expression disrupt and thereby should mutagenize the host gene, whereas enhancer trap integrations do not necessarily occur within the gene that confers transcriptional activation on the minimal promoter. This makes it very likely that only in a subset of enhancer trap integrations the endogenous gene will be mutated. For an analysis of the phenotypes caused by disruption of the endogenous genes, ES cell lines carrying reporter genes can transmit these through the mouse germline after chimera formation and appropriate breeding (6–8, 13, 14).

All three types of integrations in ES cells have been molecularly analysed to some extent. Endogenous genes have been cloned from fusion transcripts generated by activation of GT (8) and from the integration sites of ET (14, 15) and PT (13) constructs. These results demonstrate that all three vector-types can be used to identify and isolate developmentally regulated genes. For the two enhancer trap insertions analysed the β-galactosidase staining did not completely reflect expression of the endogenous gene, whereas in the two GT insertions studied the *lacZ* staining pattern and expression pattern of the endogenous gene were very similar.

Table 2. Summary and description of gene, promoter, and enhancer trap constructs

#	Construct	Type	Reporter gene elements			Selectable marker elements			Cloning aids
			Promoter/ splice acc.	Reporter	Source of pA signal	Promoter	Marker	Source of pA signal	
1	HSV-tk-lacZ	ET	ΔHSV-TK 250 bp	lacZ	SV40	No selectable marker			
2	p3LSN	ET	Δhsp68 450 bp	lacZ	SV40	HSV-TK	neo	SV40	supF
3	p6LSN	ET	Δhsp68 250 bp	lacZ	SV40	HSV-TK	neo	SV40	supF
4	Enhsrl	ET RV	Δ of enhancer of LTR of MMLV	lacZ reverse orientation	•	No selectable marker			
5	pA10neo	ET	SV40 early promoter	neo	SV40	SV40 early promoter	neo	SV40	
6	pΔE/NEO	ET RV	Δ of enhancer in LTR of MMLV	neo	MMLV U3	No additional selection marker			
7	pGT 4.5	GT	en-2 (+ 1.6 kb en-2 intron)	lacZ	SV40	Human β-actin	neo	En-2	
8	pSAβGal	GT*	Adeno major late SA	lacZ ATG (Kozak)	bgh	PGK	neo	bgh	
9	pSAβgeo	GT* lacZ/neo fusion	Adeno major late SA	lacZ			neo	bgh	
10	AcLac	GT RV	MMLV env SA	lacZ reverse orientation	•	No selectable marker			
11	pSAS	GT RV	MMLV env SA	lacZ ATG	SV40	No selectable marker			
12	ROSAβgal	GT* RV	Adeno major late SA	lacZ reverse orientation	bgh	PGK	neo	bgh	supF in LTR
13	ROSAβgeo	GT* RV lacZ/neo fusion	Adeno major late SA	lacZ reverse orientation			neo	bgh	supF in LTR
14	pT1	GT	En-2	lacZ	SV40	PGK	neo	PGK	
15	pT1 ATG	GT	En-2	lacZ	SV40	PGK	neo	PGK	
16	pT10	GT	c-fos SA 0.5 kb (exon 2)	lacZ	•	HSV-TK	neo	•	
17	pGTi	GT	MMLV env SA (+ 1.1 kb en-2 intron)	lacZ ATG (Kozak)	SV40	Human β-actin	neo	SV40	on, amp
18	pTfu	GT lacZ/neo fusion	c-fos SA 0.5 kb (exon 2)	lacZ			neo	SV40	
19	U3LacZ	PT RV		lacZ ATG	MMLV U3	HSV TK	neo	MMLV U3	
20	U3HisD	PT RV		hisD	MMLV U3	HSV TK	neo	MMLV U3	
21	U3neo	PT RV		neo	MMLV U3	No additional selection marker			

RV: retroviral construct, hsp: heat shock promoter, en-2: engrailed-2 intron sequences and splice acceptor, bgh: bovine growth hormone gene, HisD: histidinol dehydrogenase, PGK: phospho-glycerate-kinase; neo: neomycin

Percentage of target cells expr. reporter gene [%] (abs. numbers)	Restricted expr. patterns in embryos [%] (abs. numbers)		Remarks	References
	White ES lines	Blue ES lines		
[25] (5/20)	Not applicable		Oocyte injection; tested in offspring of transgenic mice	(5)
[13] (6/47)	[30] (12/41)	[50] (3/6)	Three of the white lines showed expression only in extra-embryonic tissues	(6, 14)
[58] (7/12)	[40] (2/5)	[43] (3/7)	Two of the white lines showed expression only in extra-embryonic tissues; two integrations were introduced into the germ-line and none resulted in an overt phenotype	(6, 14)
[0.5–0.6]	Not applicable		Fetal liver cells were infected and reporter gene expressing cells identified by fluorescence activated cell sorting	(39)
[100]	Not applicable		Tested in P19 EC cells; 27% (8/30) of neo^R lines turned the *neo* gene off after differentiation	(18)
[> 60]	Not applicable		Tested in rat2 cells reporter gene expressing cells identified by G418 resistance; used for bone marrow cells	Sablitzky, personal communication
[1–2] (10/600)	Not analysed	[14] (1/7)	Three cell lines showed widespread, three ubiquitous expression in embryos; three integrations were introduced into the germline and two resulted in a recessive embryonic lethal mutation	(6) (8)
[4, 6] (16/350)	[100] (1/1)	(0/2)	The two blue lines showed widespread expression in embryos; the white line showed staining following differentiation in embryoid bodies; three integrations were introduced into the germline and one resulted in a recessive embryonic lethal mutation	(7)
[95] (19/20)	[100] (1/1)	[100] (1/1)	13 amino acids between *lacZ* and *neo* coding sequences; the white line was *neo* resistant but no staining detectable in ES cells; two integrations were introduced into the germline and one resulted in a recessive embryonic lethal mutation	(7)
[0.1–0.2]	Not applicable		Fetal liver cells were infected and reporter gene expressing cells identified by fluorescence activated cell sorting	(39)
			of regulatory elements of LTR	
[4.1]	Not applicable		Tested in Cos[7]/NIH 3T3 cells, Δ of enhancer and promoter, suicide vector	(24)
[11.6] (31/268)	Not analysed	Not analysed	Δ of regulatory elements from LTRs of MMLV	(7)
[95.6] (196/205)		[27] (6/22)	Δ of regulatory elements from LTRs of MMLV; 16 cell lines showed widespread expression; 32 integrations were introduced into the germline and 14 resulted in a recessive embryonic lethal mutation	(7), Friedrich and Soriano, personal communication
[1] (300/~ 30 000)	Not analysed	[13] (35/289)	~ 35% showed widespread expression in chimeric embryos; ~ 45% no staining	Wurst *et al.* in preparation
[4] (490/11 200)	Not analysed	Not analysed		Wurst and Joyner, personal communication
[6–8]	Not analysed			Rijkers and Rüther, personal communication
[5]	(0/14)	[22] (4/18)	14 'blue' cell lines showed no expression in embryos	Zachgo, Schuster-Gossler, and Gossler unpublished
[100] (27/27)		[46] (6/13)	Two cell lines showed no expression, the remaining lines ubiquitous, widespread, or extra-embryonic expression	Soininen, Schuster-Gossler, and Gossler unpublished
[0.4]	Not applicable		Δ of regulatory elements from LTRs of MMLV; lacZ coding sequences placed into U3 LTR; tested on NIH 3T3 cells	(12)
	Not applicable		Δ of regulatory elements from LTRs of MMLV; hisD coding sequences placed into U3 LTR; *neo* present as an additional selectable marker; one integration was introduced into the germline; resulted in a recessive embryonic lethal mutation	(10, 11, 13)
	Not applicable		Δ of regulatory elements from LTRs of MMLV; *neo* coding sequences placed into U3 LTR; three integrations were introduced into the germline and one resulted in a recessive embryonic lethal mutation	(13)

phosphotransferase, HSV TK: herpes simplex virus thymidine kinase gene promoter, MMLV: Moloney Murine Leukaemia Virus, *: called promoter trap by authors, ●: information not available, Δ: deletion.

Germline chimeras have been produced by two groups with ES cell lines containing GT integrations, and recessive lethal phenotypes were found in nine out of 24 (7) and two out of three cases (8), respectively. Four PT integrations have been passed through the germline of which two resulted in embryonic lethality when bred to homozygosity (13). Only two ET integrations have been analysed after germline transmission in homozygous offspring (6, 14), and neither showed a phenotype during development. The small number of ET insertions analysed does not allow any conclusion to be drawn about the mutation rate.

In summary, cloning of endogenous genes detected by GT or PT constructs promises to be more straightforward than cloning genes from ET integration sites. Exceptions might be GT and particularly PT fusions to endogenous exon sequences which are too short to provide exon probes for further analysis. Furthermore, GT and PT integrations very likely disrupt the endogenous gene and this can be expected to cause mutations in many cases. Exceptions could include integrations which result in *lacZ* fusions to functional gene products, and GT vectors that do not inhibit splicing to downstream endogenous exons. In contrast, it is possible that few ET integrations cause mutations. Easy access to endogenous genes and a high probability of mutagenesis argue in favour of GT and PT constructs rather than enhancer traps. The higher frequency of activation of GT vectors can be an advantage whereas the PT vectors may cause more severe mutations since integrations occur in exons. However, the use of GT and PT vectors is presently limited to genes active in undifferentiated stem cells. The ET is therefore the favourable vector-type when a large number of integrations not active in ES cells is desired. Also, the higher frequency of patterns obtained with ET integrations is an advantage when the experimental aim is primarily to generate lineage and/or cell type-specific markers during development.

2. Vector design

Gene, promoter, and enhancer trap vectors to be introduced into cells consist basically of two elements: a reporter gene and a selectable marker. The *lacZ* gene of *E. coli* is the reporter gene of choice since its gene product, β-galactosidase, allows easy and sensitive *in situ* monitoring of *lacZ* gene expression in cells. The *neo* gene is a good choice of a selectable marker and can be under the control of its own promoter to achieve expression independent of the *lacZ* gene. The elements of enhancer, promoter, and gene trap vectors, and modifications of these vector-types are addressed below. A number of constructs that have been used are listed and briefly described in *Table 2*.

2.1 Basic ET constructs and minimal promoters

An enhancer trap construct in its basic version contains the *lacZ* gene with an ATG codon and polyadenylation signal under the control of a weak (minimal) promoter. A marker gene can be placed 3′ to the *lacZ* gene, to allow for selection of cells which have taken up and integrated the vector during the gene transfer (see *Figure 1A*).

The minimal promoter should only provide the sequences necessary for start of transcription without by itself being able to initiate or direct reporter gene expression. Various promoters known to have no developmental or tissue-specific activity have been used (see *Table 2*). 'Minimal' promoters have been generated by deleting most of the sequences not required for start of transcription and leaving behind a promoter which requires *cis*-acting sequences to be efficiently active. The choice of the minimal promoter is likely to influence the range of *cis*-acting sequences that are detected: some regulatory elements might act on TATA box containing promoters, while others may not.

Two fragments of the mouse *hsp 68* gene promoter (16) have been employed for ET vectors in ES cells. The vectors contained either about 450 or 200 base pairs of the promoter, (p3LSN and p6LSN respectively, see *Table 2*, #2 and 3). p3LSN contained one heat shock element upstream of the TATA box and the translation initiation codon, whereas in p6LSN this heat shock element was deleted. When integrations of p3LSN and p6LSN into the genome of ES cells were analysed for β-galactosidase activity, no major differences were found with respect to the frequency or quality of the β-galactosidase patterns that were observed in chimeric embryos produced with these cells lines (6, 14).

In a similar screen in transgenic mice (5) (in this case the ET construct did not contain a selector gene and was introduced into the mouse genome by zygote injection) 250 base pairs of the promoter region of the *HSV-tk* gene (17) were used (*Table 2*, #1). With this construct, β-galactosidase patterns were observed at comparable rates to the p3/6LSN vectors (see *Table 2* for comparison). The diversity of the patterns obtained with both of these constructs (and the fact that intact reporter gene integrations do not give rise to detectable *lacZ* expression in 'white lines') suggest that expression from these minimal promoter elements is under the control of endogenous regulatory elements.

The promoter element from the LTR of Mo-MuLV (the enhancer element was deleted) was placed in front of the *neo* gene in a retroviral ET vector (Sablitzky, personal communication, *Table 2*, #6, and below). In the absence of its own enhancer element this sequence functions as a minimal promoter and *neo* expression is controlled by endogenous regulatory elements.

To detect genes specifically expressed in stem cells, Bhat *et al.* (18) introduced into P19 EC cells an ET vector (*Table 2*, #5) that carried the SV40 early promoter in front of the *neo* gene. Previous observations demonstrated that *neo* expression from the SV40 early promoter was enhancer dependent and the endogenous regulatory elements that drove *neo* expression could be identified (18–20).

2.2 Basic GT constructs and splice acceptor sequences

A GT vector in its most basic version contains a splice acceptor sequence linked to the 5′ end of the reporter (*lacZ*) gene carrying a polyadenylation signal. As in the ET vector, a selectable marker can be placed 3′ to the *lacZ* gene (see *Figure 1B*). Functional β-galactosidase is only obtained when the integration occurs in an intron in the right orientation and when the reading frame between upstream exon(s) and the *lacZ* gene is maintained during splicing.

The only sequences needed for splicing seem to be short consensus sequences at the exon/intron and intron/exon boundaries (splice donor and splice acceptor, respectively) as well as a branch site located about 30 base pairs upstream of the splice acceptor sequence. Thus, the presence of a branch site and a splice acceptor in a gene trap vector should be sufficient for splicing if a splice donor is provided by an endogenous exon/intron boundary 5′ to the GT integration. However, the length of the sequences 5′ to the splice acceptor may influence the efficiency at which GT integrations can give rise to functional β-galactosidase. Comparing the efficiency of GT constructs carrying the mouse *En-2* splice acceptor in the context of either about 0.2 or 1.8 kilobases of intron sequences, it was found that only the construct carrying the long intron sequence gave rise to functional β-galactosidase (W. Skarnes and A. Joyner, unpublished data).

The results obtained with the two constructs mentioned above could be explained either by the 0.2 kilobase vector missing sequences required for splicing, or by our observation that in a considerable number of cases DNA integrated into ES cells, following electroporation, lose sequences from either end. Thus, the GT construct harbouring the short *En-2* intron piece may suffer frequent deletions of sequences required for splicing. The latter explanation is supported by the results with the effective GT vectors used by Friedrich and Soriano (7) that contain a 120 base pair splice acceptor site flanked upstream by retrovirus sequences. The retroviral mode of integration would in this case safeguard the vector sequences from deletions.

Splice acceptor sequences of various genes and of different lengths, (e.g. mouse *En-2, c-fos, adenovirus major late* gene, see *Table 2 #7–18*) have been used in GT constructs and have given rise to fusion transcripts and functional β-galactosidase. This suggests that any functional splice acceptor sequence should serve the purpose in GT constructs. However, the efficiency of splicing to *lacZ* versus to the endogenous downstream exon may vary between splice acceptor sequences. This of course has implications for the fidelity of the *lacZ* expression pattern and kind of mutation generated.

2.3 Basic PT constructs

The basic PT vector is identical to the gene trap vector with the exception that it does not carry a splice acceptor sequence 5′ to the reporter gene (see *Figure 1C*).

A functional reporter gene product can only be expected if the trap vector integrates into an exon of a gene and maintains the correct reading frame of the endogenous sequences with respect to the reporter gene. This event leads to a fusion transcript and a fusion protein as in the case of productive GT integrations. Integrations in 5′ untranslated exons of a gene with vectors containing an ATG are reading frame independent and generate no fusion protein.

Thus far, all tested PT constructs were based on retroviral vectors and carried in the 3′ LTR as a marker gene either the *lacZ* (U3lacZ, (12); #19 in *Table 2*), the *hisD* (U3hisD, (10, 11); #20 in *Table 2*), or the *neo* gene (U3neo, (13); #21 in *Table 2*). In the case of U3lacZ an SDK (*S*hine–*D*algarno and *K*ozak sequences) oligonucleotide in front of the *lacZ* gene provided a start-of-translation codon such that β-galactosidase expression could be expected even if no endogenous AUG preceded the *lacZ* initiation codon. The *neo* gene driven by the *tk* promoter within the vector was used to select for retroviral integration.

2.3 Selectable markers and additional sequences

2.4.1 Selectable marker

A number of marker genes have been used to select for the insertion of constructs. Dominant marker genes encoding a selectable drug resistance protein such as the *neomycin phosphotransferase* gene (*neo*) or *histidinol dehydrogenase* gene (*hisD*) are generally applicable. Other selectable markers such as the *thymidine kinase* gene (*tk*) work only in target cells with defined genotypes (see Chapter 1). Minimal requirements for the promoter that drives marker expression are that it has to be active in the target cells and should exert no regulatory effects, (e.g. enhancer properties) on the reporter gene or endogenous sequences (for examples see *Table 2*). However, even constitutive promoter elements are subject to position effects. For example, a number of different *neo* cassettes have been tested in ES cells (discussed in Chapters 1 and 2) and as much as ten-fold difference in the number of *neo*^{R} colonies was obtained. Promoter strength and selection pressure might thus be a form of pre-selection for integrations into chromosomal regions that are permissive for selector gene expression. Constructs containing fusions between the reporter and selectable marker genes (see below) or cases in which the *neo* gene serves as the (selectable) reporter gene of course do not carry functional promoters for the selector gene.

2.4.2 Cloning aids

For easy and efficient cloning it may prove valuable to introduce sequences into the trap vector that allow direct recovery of the genomic flanking sequences in bacteria. This might even be useful in the case of GT vectors if the endogenous portion of the fusion transript is too small for direct isolation of cDNA clones or for cases where it is desirable to clone the integration site. Suitable cloning aids are:

- origins of replication of plasmids in conjunction with a bacterial antibiotic resistance gene
- *supF*, a suppressor of amber mutations

When a plasmid origin of replication and an antibiotic resistance gene are inserted into the construct, sequences from the integration site can be cloned by plasmid rescue. For this, genomic DNA is digested with an enzyme which generates DNA fragments that contain genomic sequences linked to the origin of replication and an antibiotic resistance gene. Subsequent re-ligation under conditions in which intramolecular ligation is favoured followed by transformation of competent bacteria should yield only resistant colonies carrying part of the trap construct and its corresponding genomic flank (3).

When an eukaryotic *SV40* origin of replication is included as well as a bacterial origin of replication and a selectable marker, as in the case of the retroviral 'shuttle' vectors of the prototype pZIP-Neo SV(X)1 (21), sequences from the site of integration can be amplified and enriched before they are cloned by plasmid rescue. Cells harbouring the shuttle vector are fused to Cos cells expressing T antigen and this triggers replication from the *SV40* origin resulting in excision and replication of closed circular DNA molecules containing the origins of replication, the marker gene, and sequences from the genomic flank. These DNA fragments are then isolated and used to transform *E. coli* under selection pressure.

When the *supF* gene (which codes for the bacterial suppressor tRNA), which suppresses amber mutations (22), is included in the construct, cloning into lambda vectors carrying amber mutations in essential genes allows for selection for DNA fragments containing the *supF* gene. Genomic DNA can therefore be digested and ligated into these lambda vectors, (e.g. EMBL3A, see (23)) and plated on bacterial supF$^-$ hosts. Only phages carrying the *supF* gene and flanking sequences should grow. However, revertants in the amber mutations can cause background problems.

Insertion of multiple restriction sites in trap constructs are helpful so that after integration into the genome, DNA fragments of suitable length can be released that contain part of the construct and flanking sequences. This is particularly useful if inverse polymerase chain reaction (PCR, see *Protocol 15*) is used for cloning of the integration site.

2.4.3 'Buffer' sequences

As already mentioned, we observed in a considerable number of cases that linearized DNA introduced into ES cells by electroporation had lost sequences from either end after integration into the genome. With ET constructs parts of the essential minimal promoter and the 5′ end of the *lacZ* gene could be deleted. GT constructs could lose the SA of the reporter gene and PT vectors the 5′ end of *lacZ*. All constructs could lose (part of) the *neo* gene from the 3′ end although these would not give rise to *neo*^{R} colonies. In GT constructs, 'buffer' sequences can in many cases be obtained by choosing larger intron sequences in front of the splice acceptor that are devoid of any regulatory activity and which could influence expression of the reporter gene. 'Buffer' sequences that do not influence *lacZ* expression could also be placed in front of ET and PT constructs. With ET vectors we avoid placing a buffer in front of the minimal promoter and instead analyse growing ES cell clones by PCR for intact 5′ ends of the constructs (see *Protocol 2*).

2.5 Modifications of basic vector types

Various modifications and alterations of the basic vectors have been designed either:

(a) to increase 'productive' integration events in general (for GT vectors multiple splice acceptors and addition of ATG in *lacZ*),
(b) to allow direct selection of *lacZ* expressing cells (reporter/selector fusion vectors), or
(c) to provide means to increase the efficiency of introducing single copies of the trap vectors into (ES) cells without gross rearrangements at the site of integration (retroviral vectors).

2.5.1 Multiple splice acceptors and ATG codon in gene trap vectors

To obtain functional β-galactosidase activity with GT vectors, integrations have to occur within a gene, and the reading frame between the *lacZ* gene and endogenous exon sequences has to be maintained. These requirements result in a low rate of integrations with detectable β-galactosidase activity. Several modifications of the basic vector-type have been used to increase the frequency of 'productive' integrations. These modifications are:

- introduction of multiple splice acceptor sites in the three reading frames with respect to the *lacZ* gene
- addition of an ATG codon in the context of a Kozak consensus sequence to the *lacZ* gene
- a combination of these two alterations

Some 'modified' gene trap vectors that have been used are listed and described in *Table 2* (#8–13, 15, 17, and 18). Multiple splice acceptor sequences (that are usually involved in differential splicing events) which potentially give rise to various splice products with different reading frames between endogenous and *lacZ* sequences may increase the frequency of 'productive' integrations, although this has not been demonstrated conclusively. Alternatively, the reporter gene can be provided with its own start codon in a context that allows efficient start of translation (Kozak consensus sequence). This should extend the range of integration events that lead to functional β-galactosidase activity to integrations into 5′ untranslated exon sequences and into introns that follow untranslated exons. The addition of an ATG in the context of a SA may increase the rate of 'blue' target cell clones three to ten times (tested with retroviral vector (24); W. Wurst and A. Joyner personal communication; P. Soriano personal communication). This modification can also be used for PT vectors. A disadvantage may lie in the possibility that the fusion transcript may only contain a few bases of endogenous 5′ untranslated sequence which does not allow for cloning the gene at the cDNA level directly. Cloning aids (as discussed before) may provide easy access to genomic flanking sequences in such cases.

2.5.2 Reporter/selector fusions

Fusion vectors between the *lacZ* and *neo* genes have been designed to allow for direct selection for 'blue' clones ((7); Soininen and Gossler unpublished). With such vectors, functional *neo* and β-galactosidase proteins will only be produced when the integration leads to the activation of the *lacZ/neo* fusion gene. Screens based on these vectors preclude the detection of loci that are turned off in the target cells but become active later on during differentiation. In one report not all of the *neo*^{R} clones were found to be β-galactosidase positive (see *Table 2* #9 and 13) indicating that less *neo* protein might be required to confer resistance to G418 than β-galactosidase protein is necessary to obtain visible staining.

Although the 3′ end of the *lacZ* gene is essential for β-galactosidase function (25), a variety of protein fusions to a specific site in the 3′ end have been found not to interfere with enzymatic activity. Thus, various fusions between the *lacZ* gene and selector genes other than *neo* can be expected to produce bifunctional fusion proteins. However the *neo* gene activity can be affected by fusions, since it has been shown that long amino-terminal fusions could result in profound differences in the enzymatic activity of the *neo* gene product (26). If the *neo* gene product is rendered less active in a particular fusion protein, this could result in a preferential recovery of integrations into genes that are expressed at high levels. However, the βgeo vector (7) seems to work efficiently.

2.5.3 Retroviral vectors

Retroviral infection as a means of introducing foreign DNA into the genome has a number of advantages. First, the trap vector integrates at each site with a defined vector structure and little rearrangement at the site of insertion. Secondly, the infection efficiency can be 100%, which is a prerequisite for some experimental systems, (e.g. haemopoietic stem cells). Although bias in target site selection for Mo-MuLV-based vectors has been shown (27) no sequence motif that predisposes host DNA for integration is known. However, integrations seem to cluster near transcribed and DNAse I hypersensitive sites. For some experiments this could be an advantage. A detailed description of the design and application of retroviral vectors is provided in Chapter 3, a number of retroviral trap vectors are listed in *Table 2*.

2.5.4 Potential other variations

It has been shown in various eukaryotic cells, that viral sequences and sequences from the 5′ leader of the *immunoglobulin heavy chain binding protein* gene (*BiP*) allow efficient initiation of translation from an internal start of translation site in a polycistronic mRNA ((28) and references therein). If this observation holds true for ES cells, these sequences could help to eliminate the constraints associated with splicing in the correct reading frame. When a sequence that allows for internal start of translation is included in GT and PT vectors upstream of *lacZ* in front of an efficient start of translation sequence, the only requirement for expression of functional β-galactosidase should be the generation of a fusion transcript. However, one should be aware that a vector as outlined above can only be used when all cell types which are to be analysed (in tissue culture, embryos, or adult tissues) are able to recognize and use this sequence efficiently.

A signal sequence allowing for internal start of translation could also be placed between the reporter and selector genes if functional protein fusions can not be obtained. Possible inhibitory effects of the fusion between reporter and selector gene activity should thus be bypassed.

3. Establishment of cell lines carrying reporter gene integrations

3.1 Introduction of reporter gene constructs into ES cells

ES cell clones carrying reporter gene constructs integrated in their genome can be easily produced by various DNA transfer techniques. We use electroporation (Chapter 2) as a standard procedure to introduce plasmid reporter gene constructs into ES cells. Retroviral vectors are introduced by infection (see Chapter 3). The *neo* gene as part of the vector allows in both cases for direct selection of cells which have integrated vector DNA. It is important:

(a) to obtain cell lines which carry a single integration site;

(b) to ensure, that these copies are still intact, i.e. that they still contain the elements required for proper function.

Both requirements are intrinsically met when retroviral vectors are used at a multiplicity of infection (m.o.i.) below one because of the mechanism of integration of the proviral DNA. When electroporation or other transfection procedures are used, precautions must be taken to meet the conditions mentioned above (see below).

Integrations at single sites are especially important in the case of enhancer trap vectors. When a staining pattern is observed if several different integration sites are present in the cell line, it is impossible to distinguish which of the copies is transcribed. If these copies can not be segregated by breeding in transgenic animals (obtained from germline chimeras) then all integrations have to be cloned and analysed to identify the endogenous gene which was originally detected by the staining pattern. This is a tedious and time-consuming task. Single integrations are less important in the case of GT and PT integrations, because the detected 'active' copy is linked to the endogenous gene by the fusion transcript, and it is unlikely that several independent integrations in one cell will give rise to fusion transcripts.

Intact reporter genes are necessary to obtain functional β-galactosidase. The following steps can be taken to ensure that the cell lines used for further analysis in chimeras carry intact copies of the construct or to increase the likelihood that reporter genes potentially can give rise to functional β-galactosidase.

(a) Integrations can be analysed by PCR using appropriate sets of primers to demonstrate that the sequences necessary for a functional *lacZ* product are still present (see Section 3.2).

(b) Only ES cell clones which express functional β-galactosidase can be selected and used for further analysis (see Section 3.3).

3.1.1 Electroporation

The conditions we use (see *Protocol 1*) usually result in the integration of one, sometimes several tandem copies of the vector into the genome at one site. Most integrations (approx. 70%) carry single copies at one site. The following electroporation protocol could be used for all types of trap vectors. General procedures for maintaining ES cells are described in Chapter 2.

Protocol 1. Electroporation of ES cells with trap constructs

1. Electroporate ES cells as described in Chapter 2 (*Protocol 8*). To ensure integration of only a single vector copy, use DNA in the range of 5–20 μg/ml of cells (depending on the size of the vector).
2. Seed electroporated cells at a density of $2.5–5 \times 10^6$ cells/90 mm dish[a] on a gelatinized plate in medium containing LIF (Chapter 2, *Protocol 3*).
3. Add selection medium after 36–48 h. Colonies of resistant cells should appear within seven to ten days.

[a] The cell concentration per plate must be adjusted depending on the vector such that not more than 200–500 *neo*^{R} colonies are obtained on each plate.

To analyse reporter gene integrations for the presence of sequences essential for reporter gene function by PCR see *Protocol 2*.

Protocol 2. PCR analysis for intact 5′ ends of integrated reporter genes

Materials
- mouth-controlled pipette
- proteinase K (200 mg/ml frozen stock)
- appropriate DNA primers
- *Taq* polymerase
- *Taq* reaction buffer
- dNTPs
- thermocycler

Method
1. Pick part of each colony with a mouth-controlled pipette or yellow tip (see Chapter 2, *Protocol 10*) and transfer into an Eppendorf tube containing 50 μl distilled water.
2. Mix and freeze at −70 °C for 10 min.
3. Thaw, add 100 μg proteinase K (5 μl of a 20 mg/ml stock), and incubate for 1.5 h at 55 °C.
4. Heat for 10 min at 97 °C, then chill on ice for 5 min.
5. Use 25 μl for PCR and continue as described in *Protocol 15* steps 4 to 6, omitting the restriction digest, re-ligation, and cutting steps.

3.1.2 Retroviral infection

Retroviral infection is easy to perform and does not require special equipment. It is brought about by contact between the viral particles and the cells.

A high titre (between 10^5–10^7 c.f.u./ml in the absence of helper virus production) of infectious particles containing the retroviral vector is preferred for most experiments. Stable transfection of the retroviral vector sequences into packaging cell lines will lead to a titre of 10^5 c.f.u./ml or higher if the transcription and packaging of the vector sequences proceeds efficiently (see Chapter 3). Transient transfections lead to titres between 10^1–10^3 c.f.u./ml which may be sufficient for some experiments.

Protocol 3. Infection of ES cells with retroviral vectors

Materials
- gelatinized tissue culture plates (Chapter 2, *Protocol 3*)
- ES cell medium (Chapter 2) containing retrovirus

Protocol 3. *Continued*

- polybrene (Sigma Cat. No. H 9268)
- LIF or BRL conditioned medium

Method

1. Plate ES cells on a gelatinized tissue culture dish at a density of 3×10^6 cells per 90 mm dish in ES cell medium supplemented with LIF or BRL conditioned medium.
2. After 24 h aspirate medium, add 5 ml fresh medium containing the retroviral particles at an m.o.i. < 1, and 5 μg/ml polybrene.
3. After overnight inoculation remove virus containing medium, add fresh medium, and culture for an additional 24 h.
4. Add the selection medium and change medium every other day. Resistant colonies become visible between seven and ten days.

3.2 Identification of *lacZ* expressing ES cell clones

In some experiments only ES cell clones which show β-galactosidase activity in undifferentiated ES cells are to be analysed, (e.g. when GT constructs are used). Because these ES cell clones usually make up only a small fraction of the clones that carry the vector (that are *neo*^{R}) it is desirable to identify these clones as early as possible to minimize cell culture work, which is costly and labour intensive. It is preferable to analyse ES cell clones before or during 'picking' of *neo*^{R} colonies and then to expand further only the desired 'positive' cell lines. Depending on the type of vector, several strategies can be used to identify *lacZ* expressing ES cell clones rapidly.

3.2.1 Direct selection

Direct selection is only applicable for *lacZ–neo* (or, more general for reporter–selector gene) fusion vectors, which give rise to a β-galactosidase–neomycin phosphotransferase fusion protein. In these cases, resistance to the drug G418 requires and indicates presence of a β-galactosidase/neo fusion protein. However, the sensitivity of detecting those two activities may vary and cells which express functional fusions (as indicated by resistance to G418) might not show β-galactosidase activity that can be visualized in the enzymatic staining reaction using X-gal. Thus, if the purpose of the experiment is to only analyse clones that express clearly detectable levels of β-galactosidase then *lacZ* expression has to be analysed in addition to the *neo* selection. However even in such cases, the use of fusion vectors provides an easy means to strongly enrich for ES cells which show detectable levels of β-galactosidase activity. Alternatively, this may allow the specific isolation of cells that show low or undetectable levels of *lacZ* expression and thus to isolate genes that are expressed at low levels in the target cells.

3.2.2 Replica plating

Replica plating is generally applicable for a variety of cultured cells and involves the generation of a filter bound copy of each *neo*^{R} colony that can be processed to detect the subset of *lacZ* positive clones. The best results were obtained using polyester filters as described in *Protocol 4* (29, 30). Depending on the pore size of the polyester cloth, more than one replica can be made from a single plate. Polyester cloth of various pore size is available from R. Cadisch and Sons, London. The replica plating procedure does not affect the ability of ES cells to colonize the germline of chimeric mice (8).

Protocol 4. Replica plating of ES cells (provided by W. Skarnes, adapted from 30)

Materials

- polyester fabric
- glass beads (about 3 mm in diameter)
- sterile forceps
- concentrated HCl
- absolute ethanol

Method

1. Cut the fabric[a] to the appropriate size circles for the plates used and cut small notches from the edges to orient the replicas after their removal.
2. Treat the filters with concentrated HCl overnight.
3. Rinse several times in PBS, several times in deionized water, and then a final rinse in absolute ethanol.
4. Dry the filters and autoclave.
5. Treat the glass beads as in steps 2–4.
6. Place the filters on the plates approximately seven days after G418 selection, at a time when the *neo*^{R} colonies are becoming visible to the eye.
7. Aspirate off the media, gently place the polyester filter on the plates using sterile alcohol-flamed forceps, and mark the position of the notches on the bottom of the plate.
8. Overlay the filter completely with a single layer of glass beads and add back the medium.
9. Culture for four days prior to removing the filter, changing the medium every day.
10. To remove the filter, aspirate medium, invert the plate, and tap off the glass beads.

Protocol 4. *Continued*

11. With a fine pair of sterile forceps, grasp the filter at a notch and then carefully peel off the filter. Any lateral movement of the filter will dislodge colonies from the plate. Often, a portion of a colony will detach from the plate but rarely will the entire colony dislodge from the plate.
12. Gently add media to plates and return to incubator for one to three days.
13. Fix and stain each replica filter with X-gal as described for attached cells in *Protocol 9* of Section 4.3.

A potentially more reliable method for replica plating ES cells (31) involves a two layer system: a nitrocellulose filter is placed on top of a 10 μm polyester filter. In this case, the nitrocellulose filter is removed and stained with X-gal and the polyester filter is left on the plates thus preventing the detachment of colonies. The *lacZ* expressing colonies may be dissected from the filters with a razor blade and placed in culture to recover the desired cell lines.

[a] Polyester fabric with a pore size of 1 μm can be used to generate a single replica of ES cells, whereas a pore size of 10 μm can be used to produce more than one replica (31).

3.2.3 *In vivo* staining

Staining for β-galactosidase activity with X-gal requires fixation of cells prior to the enzymatic assay. Living cells can be analysed for β-galactosidase activity using fluorescein diβ-D galactopyranoside (FDG) or 'Imagene', an FDG derivative (Molecular Probes). FDG is introduced into the cells by hypotonic loading while Imagene passes freely across cell membranes but after galactosidase cleavage becomes highly hydrophobic and does not diffuse outside of the cell. Positive cells can be identified by the fluorescence of the metabolized substrate.

Protocol 5. *In vivo* staining of ES cell colonies (protocol provided by Greg Barsh, adapted from 32)

Materials

- FDG or Imagene (Molecular Probes)
- loading medium: dilute FDG stock solution (20 mM FDG in 10% DMSO) 1:10 with sterile water, a 1:1 mixture of tissue culture medium and the 1:10 diluted FDG solution is used as loading medium. (FDG stocks may vary depending on batch and supplier)
- fluorescence microscope
- ES cells grown on tissue culture dishes that fit between the stage and the objective of the microscope

Method

1. FDG: aspirate tissue culture medium and add sufficient loading medium to cover the cells, (i.e. 1 ml/35 mm dish, 2 ml/60 mm dish, 3 ml/90 mm dish). Incubate for 1 min. Change back to regular tissue culture medium. Imagene: add dye directly to the medium at a final concentration of 33 μM.
2. Incubate at 37°C for one (FDG) or two (Imagene) hours.
3. Identify fluorescing colonies with a fluorescence microscope using 10 × or 20 × objectives and filters for fluorescein. Mark the positions of positive clones[a] with a dot on the bottom of the dish.

[a] Verify positive clones by X-gal staining. The fluorescence signal of a true positive clone can vary tremendously and most of the signals localize to just a portion of the colony. Use a cell line known to express *lacZ* at clearly detectable levels as a positive control.

4. Production and analysis of chimeras

4.1 Production of chimeras

Chimeric embryos can be produced either by blastocyst or morula injection, or aggregation between ES cell aggregates and morula stage embryos as described in Chapters 4 and 5, respectively. Since β-galactosidase staining in early embryos up to about day 12 does not produce endogenous background under the staining conditions given below, any β-galactosidase staining patterns observed in embryos obtained after chimera formation are the result of expression of the reporter gene. In cases in which no staining is observed in embryos obtained after blastocyst injection or aggregation, a control is required to distinguish whether the lack of staining is due to no activation of the reporter gene or due to a lack of (or too low) contribution of ES cells in the embryos. To monitor the successful generation of chimeras and to estimate the degree of ES cell contribution in chimeric embryos a combination of ES cells derived from a pigmented mouse strain with host blastocysts derived from albino mice should be used, which allows detection of ES cell contribution to the embryo from about day 12 onwards by analysing eye pigmentation. Any pigmented cells in the eyes are ES cell-derived and therefore are indicative of the ES cell contribution to the embryo. For analysis of earlier stages we proceed according to *Protocol 6* (see also *Figure 3*).

Protocol 6. Experimental set-up to control successful generation of chimeric embryos

1. Inject 10–15 albino-derived blastocysts per early stage to be analysed with ES cells.
2. Inject an additional 10–15 blastocysts for the day 12.5 control.

Protocol 6. *Continued*

3. Transfer the embryos to foster females in groups of 10–15 (one foster per stage to be analysed and one additional for the control).

4. Recover embryos at desired stages (see *Protocol 7*), and stain for β-galactosidase activity (see *Protocol 9*).

5. Recover control embryos at day 12.5, monitor the embryos for pigmented cells in the eye (which can be easily seen by naked eye or under a dissecting microscope), and stain for β-galactosidase activity.

When a given percentage of embryos at day 12.5 are chimeric it can be assumed that the same was true for the embryos obtained from the same pool of injected blastocysts at earlier stages. When no staining is observed at earlier stages only experiments which had at least 30–50% chimeras with significant ES cell contribution in day 12.5 embryos, and for which at least 15 embryos at each earlier stage were analysed, can be taken as indicating a lack of *lacZ* expression.

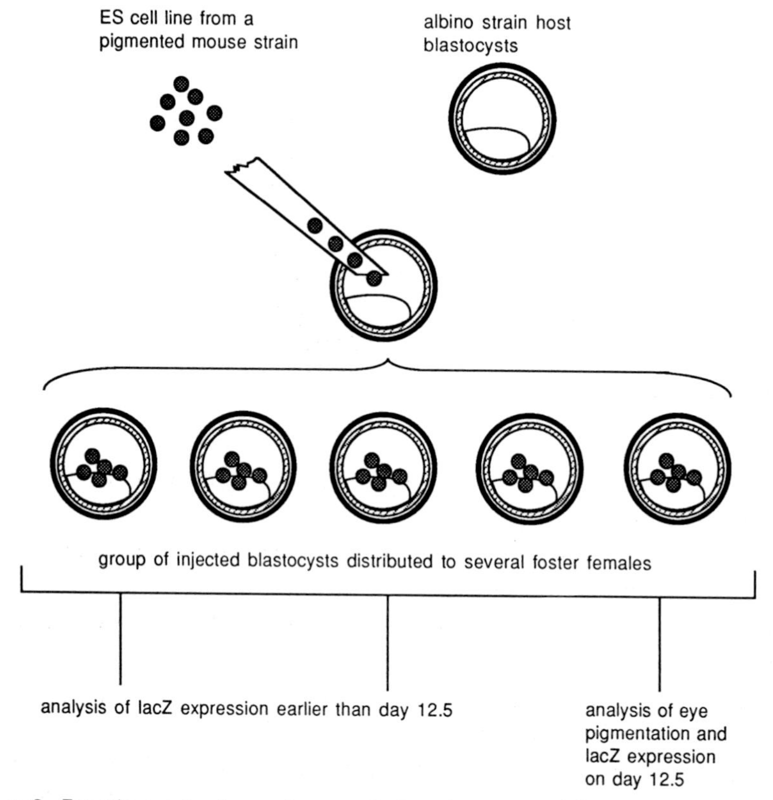

Figure 3. Experimental scheme for monitoring the successful generation of chimeric embryos.

4.2 Recovery of post-implantation embryos

Embryos obtained from injected blastocysts or aggregated morulae have to be dissected out of the uterus and away from the maternal decidual tissue before β-galactosidase staining to allow for penetration of the substrate. Since embryos are not kept in culture after recovery, no precautions for sterility have to be taken, nor has special care to be taken that chemicals used for the buffer required during dissection are of extreme purity to avoid toxic effects on the embryos. The procedure described here does not take special care to maintain the integrity of the extra-embryonic membranes, although those usually stay intact during dissection. Described in some detail is the recovery of embryos between day 6.5 and 9.5, stages at which the embryo is small compared to the decidual tissue surrounding the embryo. Later stages can be recovered more easily, the first steps being essentially identical to the procedure described.

Protocol 7. Recovery of early post-implantation embryos

Materials

- one pair of coarse forceps
- one pair of fine scissors
- two pairs of fine forceps
- two pairs of watchmakers forceps (Dumont No. 5)
- Pasteur pipettes
- Pasteur pipettes with 'wide opening' (see *Protocol 8*)
- PBS (Chapter 2, Section 2.2)
- Petri dishes
- dissecting microscope

Method

1. Kill the pregnant female either by cervical dislocation or with an overdose of CO_2.
2. Lay the mouse on its back, make a large V-shaped incision into the skin with the tip of the V just anterior to the vagina.
3. Fold the skin back, cut the abdominal wall similarly to the skin, and fold back.
4. Find the uterus, hold the uterine horns with coarse forceps at the cervical end, and cut off at the uterine–cervical junction.
5. Pull uterine horns slightly, trim away the mesometrium (part of the broad ligament that is attached to one side of each uterine horn) along the uterine wall, and cut off the uterine horns at their anterior ends.

Protocol 7. *Continued*

6. Put uterine horns into a Petri dish containing PBS.

7. Cut the uterine horns between the decidual swellings.

8. Under the dissecting microscope tear the uterine muscle at the anti-mesometrial side (side opposite to the attached mesometrium) with fine forceps.

9. Free the decidual swelling from the uterine wall by holding the muscle with one pair of forceps, and sliding along between the torn muscle and decidua with the second pair of forceps (see *Figure 4a, b*).

10. Transfer deciduas into a fresh dish containing PBS.

11. Hold the decidua at the mesometrial (broad) end with one pair of forceps. Insert the point of the closed second pair of forceps in the midline above the reddish streak (which is the embryo) through the decidua, and open forceps splitting the decidua (see *Figure 4c, d*).

12. Grasp the split parts and pull apart. The embryo usually remains attached to one decidual half.

13. Gently push the embryo with the tip of the closed forceps until the embryo is entirely free.

14. Transfer the embryo with a Pasteur pipette (days 6.5 and 7.5) or a 'wide opening' Pasteur pipette (see *Protocol 11*) (days 8.5 and 9.5). Older specimens can be transferred by 'scooping' them with curved forceps.

Removal of Reichert's membrane (days 6.5–8.5)

15. Grasp Reichert's membrane at the extra-embryonic portion of the egg cylinder with both watchmaker forceps and tear it open (see *Figure 4e, f*). Most of the membrane can be torn off leaving behind some remnants at the ectoplacental cone which do not impair staining and analysis of the embryo.

Removal of extra-embryonic membranes (> day 8.5)

16. Grasp visceral yolk sac with both forceps. Tear until the embryo is freed from the yolk sac but still connected with it by the umbilical cord. Hold the cord with one pair of forceps and tear off the yolk sac distally with second pair of forceps. If the amnion, a very thin cellular membrane, is still surrounding the embryo remove that analogously to the yolk sac.

Protocol 8. Preparation of 'wide opening' Pasteur pipettes

Materials

- diamond scratcher
- Bunsen burner
- Pasteur pipettes

Method

1. Scratch Pasteur pipette in the conical region where the diameter is about 3–4 mm with a diamond scratcher.
2. Break tip off.
3. Flame-polish the edges of the opening by heating it in a Bunsen burner until the edges start to melt.

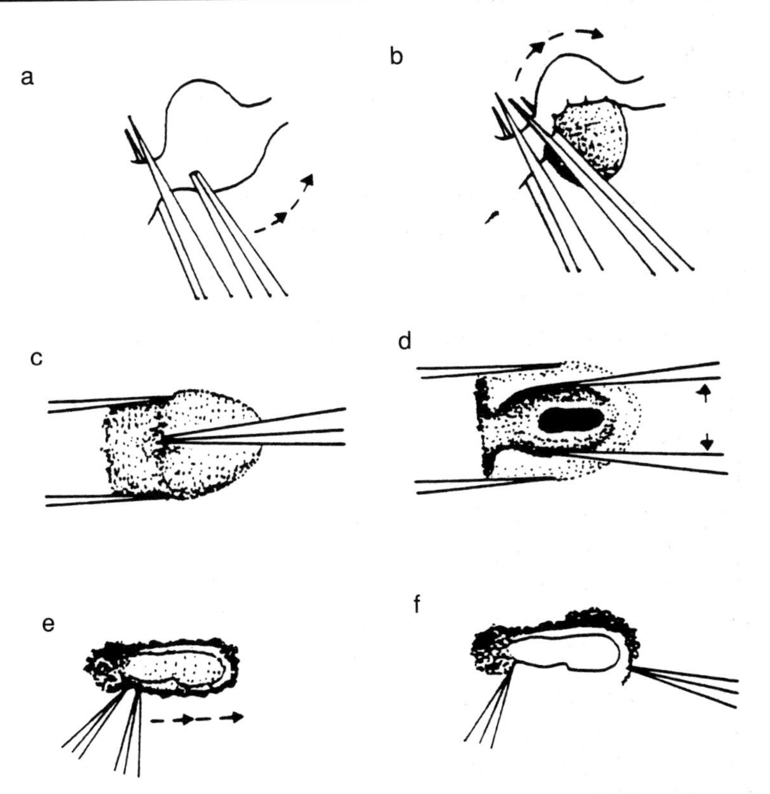

Figure 4. Diagram depicting the dissection of early mouse post-implantation embryos (taken from R. Beddington: Isolation, culture, and manipulation of post-implantation mouse embryos. In *Mammalian Development: A Practical Approach*. IRL Press, Oxford). For details see *Protocol 7*.

4.3 β-galactosidase staining

LacZ gene expression is detected by the enzymatic activity of the gene product β-galactosidase. This enzyme is quite stable and remains active even after fixation in 4% paraformaldehyde (PFA) or after freezing and thawing, which allows easy analysis of cells attached to a dish, whole embryos and tissues, as well as analysis of frozen sections. When early embryos are to be analysed then staining of whole embryos is the method of choice to analyse *lacZ* expression because it:

(a) is fast and easy to perform;
(b) allows a rapid assessment of most aspects of the expression patterns since early embryos are almost transparent;
(c) permits subsequent processing for histological examination to determine cell and tissue types expressing *lacZ* more closely.

Staining whole embryos and tissues is limited by the size of the specimen. Embryos up to day 12 can be stained as whole mounts. Older embryos tend to give staining problems either due to impaired penetration of the substrate or due to endogenous β-galactosidase activity. Therefore, in these cases either stain dissected tissues of embryos or stain frozen sections of embryos. The solutions and reagents for staining are identical in all cases.

Background staining (endogenous β-galactosidase) may be encountered in embryos from day 12.5 onwards. In most cases this is due either to:

- insufficient fixation (too short, too little volume) or
- a pH too acidic for the reaction (buffer capacity not sufficient, too little volume)

Fixation in 4% PFA is more efficient in reducing background than glutaraldehyde (GDA), but might also affect the bacterial enzyme. Using Tris-buffered saline as a basis for the staining solution reduces background but in our hands also decreases the sensitivity of the staining reaction. In cases where endogenous background causes serious problems, antibodies against the bacterial β-galactosidase and an indirect immunohistochemical assay could be tried to detect and distinguish a 'signal' (the *E. coli* enzyme) from background (endogenous enzyme).

4.3.1 Staining of attached cells, embryos, and tissues

Cells attached to the dish, embryos, and tissue samples can be stained without any special treatment prior to the procedure given below. Depending on the level of expression of the *lacZ* gene, staining becomes visible by eye after incubation for as little as 15 minutes to as long as overnight. Staining times of longer than 24 hours do not appear to increase the sensitivity.

Protocol 9. β-galactosidase staining of attached cells, embryos, and tissues

Solutions

- solution A: KPP (100 mM potassium phosphate buffer, pH 7.4)
- solution B[a]: 'fix': 0.2% glutaraldehyde (GDA) in solution A containing 5 mM EGTA and 2 mM $MgCl_2$

- solution C: 'wash': 0.01% Na desoxycholate and 0.02% Nonidet P-40 in solution A containing 5 mM EGTA and 2 mM $MgCl_2$
- solution D[b,c]: 'stain': 0.5 mg/ml X-gal[d], 10 mM $K_3[Fe(CN)_6]$[e] and 10 mM $K_4[Fe(CN)_6]$[f] in solution C

[a] Always prepare fresh prior to use. Other fixations can also be used, e.g. 2% GDA or 4% PFA are possible.
[b] When prepared freshly, chill on ice for 10 min, and spin down precipitate.
[c] Solution D can be reused several times. Filter after each use and keep in dark at 4°C in between.
[d] X-gal stock solution can be made at 50 mg/ml X-gal in dimethylsulfoxide or dimethylformamide (DMF). Keep at −20°C in dark. DMF stays liquid at −20°C.
[e,f] $K_3[Fe(CN)_6]$ and $K_4[Fe(CN)_6]$ are kept as 0.5 M stock solutions in dark bottles.

Method

1. Wash in PBS.
- for cells: aspirate the medium and replace with PBS; repeat
- for embryos: transfer into PBS, gently swirl around; repeat
- for tissues: transfer into PBS, gently swirl around; repeat

2. Fix in solution B.
- for cells: add sufficient solution B to the plate such that the cells are well covered and leave for 5 min. at rt
- for embryos: up to day 9.5 add 1 ml solution B for 10–20 embryos[a] and leave for 5 min at rt; for day 10.5 to 12.5 embryos[b] add 5–10 ml solution B for ten embryos and leave for 15 min. at rt
- for tissues: add about ten times the volume of the sample of solution B and leave 15 to 60 min. (depending on size) at rt

3. Wash three times with solution C at room temperature.
- for cells: 5 min. each
- for embryos: up to day 9.5, 5 min. each
  day 10.5 to 12.5, 15 min. each
- for tissues: 15 to 30 min. (depending on size)

4. Replace solution C with solution D and incubate at 37°C.
- for cells: add sufficient solution to the plate that cells are well covered
- for embryos: up to day 9.5 add 1 ml solution for 10–20 embryos
  day 10.5 to 12.5 add 5–10 ml for ten embryos
- for tissues: add about ten times the volume of the sample

5. After staining (1–24 h) wash samples in solution C.

Samples can be stored for a few days in solution C at 4°C, but for prolonged storage the specimens should be fixed in 4% PFA for 2 h and kept in 70% ethanol at 4°C.

[a] Convenient for staining are 30 mm (or smaller) Petri dishes.
[b] Convenient for staining are 60 mm Petri dishes.

4.3.2 Staining of cryostat sections

Staining frozen sections is required when older embryos or large tissues are to be analysed. In addition, frozen sections are desirable when tissues are to be analysed for β-galactosidase activity and for a second marker, (e.g. with antibodies, to identify particular cell types in the sample). Cryostat sections can be made from freshly-frozen material or from tissue which was frozen and stored at −70°C. When material is stored at −70°C care should be taken that samples do not dry out, (e.g. store in screw cap tubes sealed with Parafilm). We have not observed loss of β-galactosidase activity in samples which were stored for up to six months under these conditions.

Gelatinized slides are used for better adhesion of sections and can be prepared in advance and stored.

Protocol 10. Gelatinization of slides

Materials

- histological staining trays
- gelatin, (e.g. Sigma G2500) solution (0.1% w/v in sterile distilled water)
- ethanol
- slides

Method

1. Submerge (wash) slides in 100% ethanol for 30 min. (in staining tray).
2. Remove slides from ethanol and air-dry.
3. Submerge slides in gelatin solution for 20 min. (in staining tray).
4. Air-dry and store protected from dust.

We routinely cut sections at 10 μm thickness. For most tissues sectioning at temperatures between −15 to −20°C is fine. Slow freezing of embryos gives better tissue morphology compared to rapid freezing in liquid nitrogen or on dry-ice. Stained sections can either be mounted directly or can be counter-stained first.

Protocol 11. Production and staining of frozen sections

Materials

- OTC embedding medium, (e.g. Tissue Tek, Miles)
- scalpel blades, coarse forceps, small spatula
- gelatinized slides (see *Protocol 10*)

- cryostat
- histological staining trays
- embedding medium, (e.g. Eukitt, RiedelDeHaen, FRG)

Method

1. Place about 0.5 ml OTC embedding medium on to the pre-cooled (−20 °C) mounting block of the cryostat (depending on the machine used several mounting blocks can be kept at the chamber temperature of the cryostat, and several specimens can be handled simultaneously). When this layer is almost frozen, add more embedding medium and the specimen which should be free of excess liquid. Excess liquid can be removed from specimens by transferring them through several drops of embedding medium at room temperature and then to the block (early embryos can be frozen in the deciduum).
2. Orientate the specimen with a small spatula while the embedding medium is still liquid. Often it is easier to first freeze the specimen (without or in a small amount of embedding medium), and then mount the frozen sample on to the block as is done with pre-frozen and stored material.
3. When the specimen is properly oriented and completely frozen add sufficient embedding medium to surround and slightly cover the sample. Avoid air bubbles.
4. Trim away excess parts of frozen embedding medium with scalpel blade leaving behind a few millimetres around the sample.
5. Fix freezing block on to the block holder of the microtome (usually the block is put on the holder and fixed by a screw) and fine adjust the angle between the specimen and knife (this option depends on the make of the cryostat).
6. Move the knife to the specimen and trim away embedding medium covering the outer edge of the sample (this can be done by setting the section thickness to 50–100 μm). When you reach the sample, or desired area of the sample, go down to 10 μm and start to collect sections.
7. Collect sections on gelatinized slides by approaching the section (which still lays on the surface of the knife) with the slide (slide at room temperature). When slide just touches the section the latter will melt on to the glass. Do not press the slide on to the section and knife! Several sections can be collected on one slide.
8. Place the slide at room temperature or on a heating plate at 37°C and allow it to dry. This will only take a few seconds. Store dry sections at room temperature while the rest of the sections are cut.
9. Fix slides in solution B (see *Protocol 9*) for 5 min. at rt (convenient are histological staining trays).

Protocol 11. *Continued*

10. Wash three times in solution C (see *Protocol 9*), 5 min. each.

11. Stain in solution D (see *Protocol 9*) overnight.[a]

12. Wash slides first for 5 min. in solution C (see *Protocol 9*), then 5 min. in distilled water. When sections are to be counterstained proceed according to *Protocol 14*.

13. Otherwise dehydrate sections in ethanol (70%, 96%, 100% EtOH 5 min. each).

14. Pass through xylene: EtOH 1:1 (1 min.) and xylene (1 min.). Do not dry, only allow excess xylene to rinse off the slide and wipe the back.

15. Place slide horizontally, pour permount (for a coverslip of 24 × 50 mm about four drops) on sections, and place the coverslip first with one end on the slide. Keep the other end up with a needle and slowly go down allowing air bubbles to escape.

[a] Histochemical staining trays are very convenient and allow easy handling of 20 slides at a time when two slides are put back to back.

4.4 Histological analysis of stained whole mount embryos and tissues

Stained embryos and tissues can be processed further for histological analysis after refixation and dehydration. Fixation and dehydration times are given in *Table 3*. Both methacrylate and paraffin embedding give good results.

Table 3. Fixation and dehydration times for embryos and tissues

	Fixation times in 4% PFA[a]	Dehydration times in EtOH[b] EtOH concentration			
Specimen		**50%**	**70%**	**96%**	**100%**
Embryos up to day 10.5	30 min	30 min	30 min	30 min	30 min
Embryos between day 11.5 and 12.5	1 h	1 h	2 h	2 h	1 h
Embryos between day 13.5 and 15.5	2 h	2 h	4 h	4 h	2 h to overnight
Embryos older than day 15.5 and whole adult tissues	overnight	2 h	2 × 4 h	overnight	overnight

[a] 4% paraformaldehyde in phosphate-buffered saline.
[b] Wash specimen after fixation and prior to dehydration for 2 h with water.

4.4.1 Methacrylate embedding and sectioning

Dehydrated samples can be directly processed further for methacrylate embedding. The protocol we give here is based on the manufacturer's recommendations with minor alterations. The procedure does not impair the blue reaction product of the *lacZ* staining. The required embedding forms can be obtained together with all reagents as a kit (Histo-Technique-Set, Kulzer, FRG). We use the disposable knives provided with the kit but any D-type knife is fine. We section at about 5–7 μm. Thinner sections down to 1 μm are possible but some staining is not visible in thin sections.

Protocol 12. Methacrylate embedding of stained material

Materials

- Petri dishes
- Technovit 7100, hardener, and Technovit 3040 (Kulzer, FRG)
- teflon embedding forms and Histoblocs (Kulzer, FRG)
- microtome
- embedding medium, (e.g. Eukitt, RiedelDeHaen, FRG)

Method

1. Incubate the specimen in pre-infiltration solution (Technovit 7100/EtOH, 1:1). Petri dishes can be used. We use the following incubation times:

	embryos			adult tissue, e.g. brain
	up to day 12.5	day 13.5 to 16.5	> than day 16.5	
incubation time	2 h	12 h	24 h	24 h

2. Place the specimen in preparation solution (1% hardener-1 (w/v) in Technovit 7100). Times are as given above.
3. Prepare embedding medium (add one part hardener-2 to 15 parts of preparation solution). Process immediately as pot life is about 4–5 min.
4. Fill mould of embedding form half with embedding medium, add and orient specimen with a needle, and fill mould up to the top. Material will polymerize in about 2 h.
5. After hardening, glue Histobloc to embedded specimen using Technovit 3040 as given in the manual. After 15 min. specimen can be removed from the mould and sectioned. The embedding material is not removed from the sections.

4.4.2 Paraffin embedding and sectioning

Any paraffin embedding protocol that avoids solvents which affect the blue staining is fine. Short times of exposure to xylene are tolerable; prolonged incubation washes out the staining. *Protocol 13* uses isopropanol for pre-infiltration, since even incubation for three to four days in isopropanol does not affect the blue staining.

Protocol 13. Paraffin embedding of β-galactosidase stained tissues

Materials

- paraffin for histology, (e.g. Histowax, Reichert and Jung, FRG)
- isopropanol
- screw cap tubes, (e.g. 50 ml Falcon tubes)
- casting mould, (e.g. Reichert-Jung, FRG; several sizes are available)
- small spatula
- microtome and holder for fixing the paraffin block to the microtome
- xylene

Method

1. Place the dehydrated sample in 100% isopropanol for 2 h with one change of the isopropanol. Alternatively, samples can be dehydrated directly in isopropanol similarly as given in *Table 3* for ethanol. (Using isopropanol the steps are 50%, 75%, 90%, and two times 100% with incubation times as in *Table 3*.)
2. Pre-infiltrate with paraffin/isopropanol (1:1) at 60°C

	embryos			adult tissue, e.g. brain
	up to day 10.5	day 11.5 to 16.5	> than day 16.5	
incubation time	2 h	12 h	24 h	24 h

3. Infiltrate with paraffin at 60°C. The times given above apply.
4. Place into pre-warmed (60°C) mould, orientate the specimen with a needle or spatula, and fill mould with paraffin.
5. Directly cast the block holder to the embedded specimen according to the manufacturer's specification.
6. Dewax sections 1–2 min. in xylene. Sections can now be embedded or processed for counterstaining.

4.4.3 Counterstains

All staining procedures which do not remove or mask the β-galactosidase staining or affect the plastic embedding can be used. We routinely counterstain with haemoxilin/eosin under conditions which give only a weak colour. When *lacZ* staining is in the nucleus eosin stain alone may be preferable. If *lacZ* staining is very faint we counterstain only every other slide. This allows for comparison of *lacZ* staining in the otherwise uncoloured sections with adjacent counterstained sections and facilitates identification of *lacZ* expressing cells. Since methacrylate sections are affected by ethanol we give a procedure for ethanol-free reagents which we use for all sections including frozen sections. Paraffin sections have to be rehydrated after dewaxing in descending ethanol concentrations (100%, 96%, 80%, 60% ethanol, distilled water, two minutes each).

Protocol 14. Counterstaining sections of *lacZ* stained material

Materials

- histological staining trays
- haemoxilin and eosin solution (for recipes see *Table 4*)
- 60%, 80%, 96%, and 100% ethanol
- xylene
- embedding medium

1. Submerge sections for 1 min. in haemoxilin solution.
2. Rinse 5 min. with tap water.
3. Submerge sections for 2 min. in eosin solution.
4. Rinse 5 min. with distilled water.
5. Dehydrate frozen sections and paraffin sections in ascending ethanol (60%, 80%, 96%, 100%, 2 min. each).
6. Remove the ethanol by two incubations in xylene (1 min. each) and embed in Eukitt or any other commercially available embedding medium. Methacrylate sections can be dried and embedded directly.

---

5. Cloning endogenous genes

5.1 ET integrations

Cloning of genes that have been detected by enhancer trap integrations require the isolation of genomic sequences flanking the integration site since

Table 4. Recipes for counterstains

Haematoxylin solution:		**Eosin solution:**	
Haematoxylin	1.5 g	Eosin (water soluble)	0.5 g
Sodium iodate	0.15 g	Glacial acetic acid	2–3 drops
Aluminium sulfate	13.2 g	Distilled water	100 ml
Distilled water	172.5 ml		
Ethylene glycol	162.5 ml		
Glacial acetic acid	15 ml		
	350 ml		100 ml

these represent the physical linkage between the staining pattern and the endogenous locus. We have employed inverse PCR and phage libraries for cloning. Depending on the size of the flanking sequences obtained in the initial cloning, more of the wild-type locus might have to be isolated in order to identify transcribed sequences in the vicinity of the integration. Once the endogenous locus has been cloned, a comparison of the integrated and the wild-type version can be carried out to determine if the insertion event led to any reorganization at the integration site.

5.1.1 Cloning of flanking sequences and wild-type loci

The copy number and number of integration sites of ET constructs can be deduced from genomic Southern blot analysis of target cell line DNA probed with vector sequences. Considerations as to whether inverse PCR or phage libraries should be employed in specific situations are briefly discussed below.

(a) Inverse PCR is the method of choice if only one copy of the ET construct has integrated into the genome. Inverse PCR allows the direct amplification of flanking sequences based on primers annealing to the integrated vector (33, 34). The flanking sequences obtained usually do not exceed 1–2 kb and thus more DNA surrounding the site of integration ('wild-type locus') usually needs to be cloned to identify and isolate transcribed sequences. To do this a genomic wild-type library of some sort (phage or cosmid) is required. Unless sequences from both sides of the insertion are isolated no direct comparison between the wild-type and the integrated version at the site of insertion is possible.

(b) The construction of phage libraries is preferable if multiple copies of the vector have integrated, or, if no fragments of suitable length containing vector sequences and genomic flanking DNA can be identified to apply inverse PCR. Two types of libraries can be made.

   i. Phage libraries produced with partially digested genomic DNA contain the integration site and the wild-type locus because the ES cell line is hemizygous for each integration. These libraries allow the cloning of the integrated and subsequently of the wild-type version.

ii. Phage libraries produced with completely digested and size selected genomic DNA enrich for a subset of the genome including fragments that contain the integration site. The wild-type allele may or may not be contained in such a library depending on the size selection. This type of library may be preferable if wild-type genomic libraries are available and only the integration site is to be cloned from the freshly prepared library.

Protocols for the preparation of both types of lambda libraries are presented in detail in (23) and (35).

i. Inverse PCR

The cloning strategy using inverse PCR is schematically outlined in *Figure 5*. Different batches of circularized DNA may work with varying efficiency

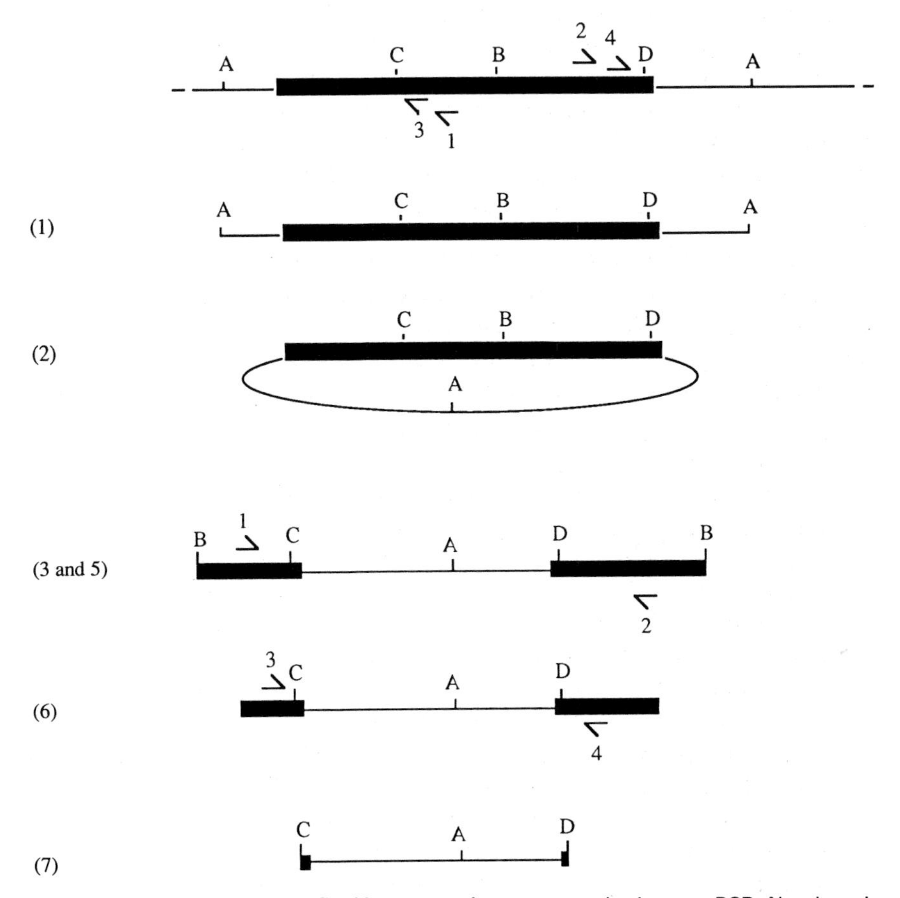

Figure 5. Principle of cloning flanking genomic sequences by inverse PCR. Numbers in brackets correspond to steps in *Protocol 15*. A–D: sites for restriction enzymes. 1–4: oligonucleotide primers. Bar: known ET vector sequences. Line: genomic sequences flanking the site of integration.

during the amplification. If the DNA from one ligation works in a test PCR then there should be sufficient material to amplify enough of the desired fragment for cloning. In our hands fragments between 1.5–2 kb have been amplified under the conditions given below. These conditions may have to be altered if larger fragments are to be amplified.

Protocol 15. Cloning flanking sequences with inverse PCR

Materials

- genomic DNA carrying the transgene insertion
- oligonucleotide primers (1 and 2 in *Figure 5*)
- restriction enzymes A–D in *Figure 5*
- T4 ligase (Boehringer, Mannheim)
- *Taq* polymerase, appropriate reagents and equipment for PCR

Method

1. Digest genomic DNA to completion with a restriction enzyme (A in *Figure 5*) that does not cut within vector DNA sequences between oligonucleotides 1 and 2 and that cuts at least 0.5–1 kb outside the vector in the genomic flanking DNA.[a] Heat inactivate (15 min, 65 °C) the restriction enzyme when digestion is complete.
2. Ligate the digested DNA with 1–2 Weiss units of T4 DNA ligase at 14 °C, for 16 h at a concentration of 1 μg DNA/ml. The dilution favours intramolecular as opposed to intermolecular ligation. Heat inactivate (15 min., 65 °C) the ligation mix.
3. Relinearize with appropriate restriction enzyme[b] (B in *Figure 5*). Add the enzyme and its 10 × incubation buffer (see manufacturers data sheet) directly to the ligation reaction. After appropriate incubation heat inactive (15 min., 65 °C) the restriction enzyme.
4. Precipitate the DNA in high salt and isopropanol in the presence of 10 μg/ml tRNA.
5. Set up the amplification reaction by combining:
   - 1–100 ng genomic DNA[c] (cut, ligated, and re-cut)
   - 5 mM dNTPs
   - 1–2 U *Taq* polymerase
   - 0.1–0.5 μM primers (1 and 2 in *Figure 5*)
   - 5 μl 10 × incubation buffer (manufacturer's recommended buffer)
   - H_2O to a total volume of 50 μl

The following conditions apply:

number of cycles	94°C (denaturation)	T_m (annealing)[d]	72°C (extension)
1	10 min[e]	2 min	2 min
30	1.5 min	2 min	2 min
1	1.5 min	2 min	10 min

6. Analyse 20 μl of each test reaction on a 1% agarose gel. A band corresponding in size to the fragment deduced from primer placement and genomic Southern blot data should be visible on ethidium bromide stained gels.[f] Confirm that the correct piece of DNA has been amplified by Southern blot analysis of the gel by hybridization with vector sequences contained in the amplified fragment.

7. Digest the rest of the test reaction, or additional reactions, with restriction enzymes which lie outside the first primer pair but within vector sequences (C and D in *Figure 5*). Alternatively PCR primers (step 5) that contain internal restriction sites can be designed. Analyse an aliquot of this digestion on an agarose gel. The amplified fragment should have shifted downward by the anticipated number of base pairs.

8. Purify the DNA fragment from an agarose gel, (e.g. by electroelution, or low melting point agarose).

9. Ligate the digested and purified PCR fragment to a plasmid vector cut with the same enzymes. If no convenient restriction enzymes (C, D; which do not cut in the genomic flank) exist, clone the eluted DNA blunt-ended into a plasmid vector.[g]

10. Transform competent bacteria with an aliquot of the ligation reaction and screen transformants for the presence of the desired clone.

[a] Shorter flanking sequences may not contain enough unique sequences to be used as a probe.
[b] Linearization of the circular DNA in the vector sequences between the primers 1 and 2 may improve the PCR amplification step.
[c] 50–100 ng of genomic input DNA worked best in our hands.
[d] As a rough estimate the primer melting temperature (T_m) = 2(X) + 4(Y); where X = Σ of dATPs + dTTPs, and Y = Σ of dGTPs + dCTPs; an applicable annealing temperature is 5°C below the true T_m.
[e] Optional without *Taq* polymerase; add enzyme before annealing step of first cycle.
[f] A second round of amplification with nested primers (3 and 4, *Figure 5*) is optional if the material obtained by the first round of PCR is not sufficient for cloning. Use 0.1–1 μl of the first PCR reaction and add new primer, dNTPs, 10 × reaction buffer, and *Taq* polymerase.
[g] Pre-treatment with T4 polymerase (36) enhances the efficiency of this step.

The efficiency of the first steps of the protocol can be estimated on agarose gels. If multiple bands are visible after the PCR step, try less input DNA or raise the annealing temperature above T_m. Primer–dimer artifacts and other factors, (e.g. buffer, dNTP concentration) influencing the specificity of amplification are discussed in detail in (37).

5.1.2 Isolation of cDNAs corresponding to transcribed sequences near the site of integration

Genomic sequences upstream and downstream of an enhancer trap integration site have to be searched for the endogenous gene(s) that was (were) detected by the enhancer trap β-galactosidase expression, since enhancers can act bidirectionally on either side of a promoter. Enhancers may be located at large distances from the promoters they are regulating, however, frequently they are found in the direct vicinity of the sequences whose transcription they control. We have analysed about 10–20 kilobases of genomic sequences on either side of two enhancer trap integrations and have identified transcripts in both cases within five kilobases of the integration site (14, 15).

Either 'unique' sequences present in the flanking DNA can be identified and can be used as probes for screening cDNA libraries (from developmental stages at which reporter gene activity has been detected), or, the total flanking DNA contained in a phage or cosmid clone may be taken as a probe. If large genomic fragments are used as probes, repetitive sequences have to be blocked with an excess of 'cold' total genomic DNA as described in *Protocol 16*. This method can be sensitive enough to find 100 base pairs of coding sequence in 40 kilobases of genomic DNA.

Prior to screening cDNA libraries, it is useful to analyse flanking sequences for probes that have been conserved during evolution. This can be accomplished by probing Southern blots of DNA digests from a variety of organisms. Conserved sequences can be indicative for the presence of coding sequences and thus may be preferable probes.

Since the genomic probes to be used may contain any part of the endogenous gene, (e.g. the very 5′ or 3′ end), random primed cDNA libraries which should contain all regions of a gene or oligo dT-primed libraries known to contain long cDNAs should be used.

Protocol 16. Preparing hybridization probes from total phage insert DNA (protocol provided by B. Herrmann)

Materials

- flanking DNA (20 ng/μl)
- random hexamer priming kit (Boehringer Mannheim)
- vector DNA for blocking reaction
- 1 M Na_2HPO_4/NaH_2PO_4 buffer pH 6.8 (NaPi)
- sonicated genomic DNA

Method

1. Boil 50–100 ng of flanking DNA (phage lambda insert) in 50 μl of TE (10 mM Tris/HCl, pH 8; 1 mM EDTA, pH 8) for 5 min. and place on ice immediately.

2. Label 20 ng of the denatured DNA by random hexamer priming (rt, several hours or overnight; Boehringer, see manufacturer's recommendations) as recommended by the manufacturer. Store the rest of the DNA at −20°C for later use.
3. Precipitate the probe in high salt and isopropanol with 5 μg of yeast tRNA for 5 min. on ice, and pellet for 5 min. in microcentrifuge.
4. Dissolve the pellet in 20 μl TE.
5. Add 100 μg of unlabelled total genomic DNA (in 156 μl H_2O) sonicated to 2 kb average size and boil for 7 min.
6. Add 24 μl of 1 M NaPi pH 6.8, incubate in a pre-heated lead block for 2 h at 65–68°C, and put on ice until use (preferably right away).
7. Boil 2–6 μg of vector DNA (if phage lambda sequences have to be blocked) in 20 μl of H_2O for 20 min.
8. Add 2.2 μl of 1 M NaPi pH 6.8 and add to adsorption reaction (step **5**) at least 30 min. before end of reaction.
9. Hybridize the cDNA library nylon filters (UV cross-linked) in 0.3 M NaPi pH 6.8 (mixed buffer system, is approximately 0.5 M in Na), 7% SDS, 5 mM EDTA at 68°C for 16 h. The concentration of the probe should be 0.5–1 $\times$ 10^6 c.p.m./ml hybridization solution; pre-hybridization is done for 1 h in the same solution.
10. Wash the filters in 40 mM NaPi pH 6.8, 1% SDS at 68°C with three or four changes of the wash solution.

The following are a number of important points to keep in mind.

(a) If the vector sequences have to be competed out, there will be a few blotches on the autoradiographs which should not affect the identification of positives on the duplicate filters.

(b) A large batch of unlabelled total genomic DNA should be made and sonicated. First try the blocked probe on a genomic Southern blot. If bands can be distinguished in the lanes, then the probe can be used to screen libraries.

(c) Hybridization and wash steps should be carried out with pre-warmed solutions and the filters should be 'coated' with SDS.

(d) Mix the probe in hybridization solution before adding it to the filters.

5.2 GT and PT integrations

For GT and PT integrations the physical linkage between the β-galactosidase activity and the endogenous gene is the *lacZ* fusion transcript. 'Blue' ES cell lines provide a convenient source of RNA for cloning these *lacZ* fusion cDNAs. The endogenous portion of the fusion transcript can be cloned by means of the rapid amplification of cDNA ends (RACE) (see *Protocol 17* and (38)) which provides an alternative to preparing cDNA libraries from each GT cell line. In the case of 'white' GT ES cell lines, cloning by the same method can be attempted from RNA isolated from chimeric embryos or, after germline transmission, from transgenic embryos, Alternatively, the endogenous genomic locus can be isolated first. This may be preferable for PT integrations since the flanking sequences must contain transcribed sequences.

5.2.1 Cloning endogenous exon sequences contained in the *lacZ* fusion transcript using the RACE protocol

The RACE protocol, with some modifications, was used to clone the endogenous genes associated with three gene trap insertions (8). The strategy used to clone sequences 5′ of *lacZ* in the fusion transcript is outlined in *Figure 6*. Briefly, first strand cDNA is primed with an oligonucleotide complementary to the 5′ end of *lacZ*. These products are purified and tailed with dATP residues. Second strand cDNA is then prepared with a second oligonucleotide primer containing rare cutting restriction enzyme sites and a dT-tail. The use of nested primers for the PCR reaction is essential to obtain the desired

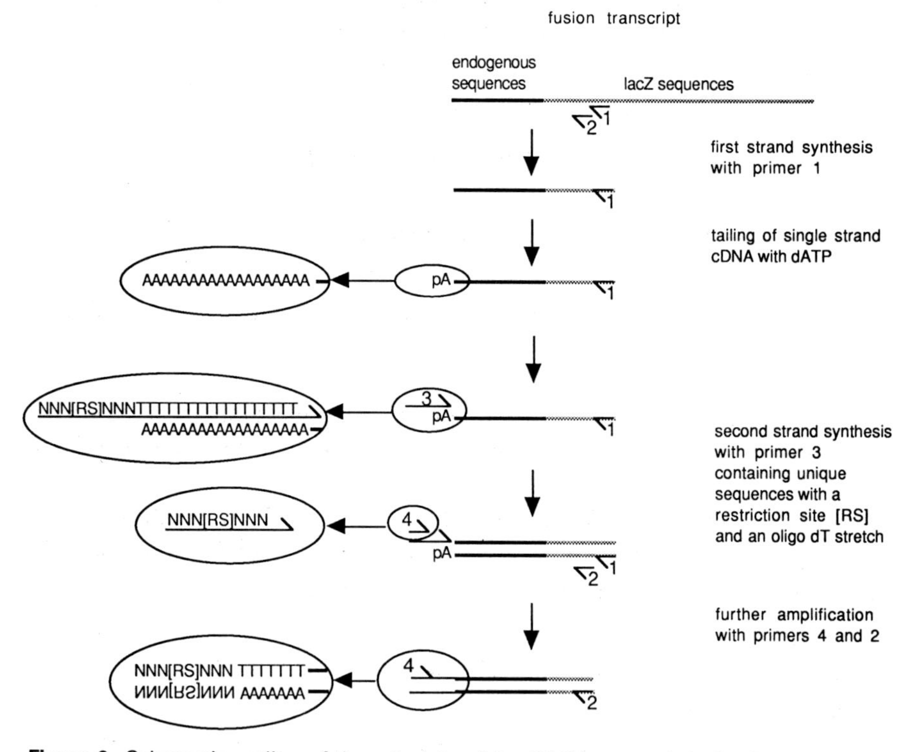

Figure 6. Schematic outline of the principle of the RACE protocol. 1–4: oligonucleotide primers. For details see *Protocol 17*.

cDNA products since the conditions used in synthesizing the first strand may be relatively non-specific. The cDNA products are thus amplified using a third oligonucleotide primer complementary to sequences upstream of the original *lacZ* primer in combination with the second primer (minus the dT-tail). The amplified cDNA is then cloned using restriction endonuclease sites present at either end of the fragments. It is advantageous to include a portion of the *lacZ* coding region in the amplified cDNA since these sequences can be used as a probe to identify the desired cDNA clones.

Protocol 17. PCR amplification of the 5′ ends of *lacZ* fusion transcripts (protocol provided by W. Skarnes)

Materials

- oligonucleotide primers (1 to 4 in *Figure 6*)
  primers that have been used in (8):
  number 1: 5′CGTCGTGACTGGAAAACCCT3′
  number 2: 5′CCAACAGGTACCTGACAGAGCA3′
  number 3: 5′GGTTGTGTCGACTATCGATGGGTTTTTTTTTTTTTTTT3′
  number 4: 5′GGTTGTGTCGACTATCGATGGG3′
- MuLV reverse transcriptase (Life Sciences)
- DEPC-treated deionized water (ddH_2O)
- NACS column (Bethesda Research Laboratories)
- RNasin (Promega)
- terminal transferase (Boehringer Mannheim)
- Klenow (Boehringer Mannheim)
- *Taq* polymerase (Cetus), appropriate reagents and equipment for PCR

Method

1. Purify total RNA from each ES cell line using guanidinium thiocyanate followed by centrifugation through a caesium chloride cushion as described (23).
2. Resuspend 5 μg total RNA in 8.5 μl double distilled water (ddH_2O), add 5 ng of oligo number 1 (*Figure 6*), and incubate at 68 °C for 5 min.
3. Cool on ice and add to a total volume of 20 μl:
   - 1 μl 20 mM dithiothreitol
   - 3 μl mixture of deoxyribonucleotides (each at 10 mM)
   - 0.25 μl RNasin
   - 2 μl 0.5 M Tris pH 7.3, 0.1 M magnesium chloride
   - 10 U MuLV reverse transcriptase

Protocol 17. *Continued*

4. Incubate at 42 °C for 2 h (first strand synthesis), then stop the reaction by adding 80 μl ddH_2O, 20 μl 10 M ammonium acetate, and 320 μl absolute ethanol.
5. Precipitate on ice for 5 min., spin in microfuge for 5 min., discard supernatant, wash pellet with 70% ethanol, and dry pellet.
6. Hydrolyse the RNA in alkali to improve the efficiency of the dA-tailing reaction by resuspending the first strand cDNA products in 17 μl ddH_2O, add 18 μl 0.2 M sodium hydroxide and incubate at 65 °C for 1 h.
7. Add 25 μl 0.1 M hydrochloric acid and 25 μl 1 M Tris–HCl pH 8.0 to neutralize.
8. Affinity purify the single stranded cDNA on a NACS column as recommended by the manufacturer.[a]
9. Carry out the dA-tailing of the first strand cDNA in 40 μl of buffer (0.14 M potassium cacodylate, 30 mM Tris–HCl pH 7.6, 1 mM cobalt chloride, 100 mM dithiothreitol), containing 0.2 mM ATP, and 50 U of terminal transferase.
10. Incubate the reaction for 7.5 min. at 37 °C, add 110 μl ddH_2O, and stop the reaction by extracting once with phenol:chloroform (1:1).
11. Extract a second time with chloroform.
12. Add 50 μl 10 M ammonium acetate and precipitate with 0.5 ml absolute ethanol.
13. Spin, wash pellet with 70% ethanol, and dry the pellet.
14. Use a small amount of dT-tailed oligo number 2 (17 T residues added to the 3′ end) to prime the second strand synthesis using the Klenow enzyme. Resuspend the dA-tailed cDNA in 15 μl of buffer (50 mM Tris–HCl pH 8.3, 10 mM magnesium chloride), containing; 1 ng dT-tailed oligo number 2, 1 mM of each deoxyribonucleotide, and 5 U of Klenow enzyme.
15. Incubate the reaction at room temperature for 30 min., and then at 37 °C for another 30 min.
16. Stop the reaction by adding 65 μl ddH_2O, 20 μl 10 M ammonium acetate, and precipitate with 250 μl absolute ethanol.
17. Spin down reaction products, wash the pellet with 70% ethanol, and dry the pellet.
18. Carry out the polymerase chain reaction with an excess (1 μg) of oligo number 2 and oligo number 3 in 50 μl of buffer (10 mM Tris–HCl pH 8.0, 50 mM potassium chloride, 1.25 mM magnesium chloride, 0.01% gelatin), containing 0.2 mM of each deoxyribonucleotide and 5 U *Taq* polymerase. For 22-mer oligos with 50% G:C content (the optimal amplification

conditions should be determined empirically for each set of primers), the reaction is carried out as given below:

number of cycles	94°C (denaturation)	T_m (annealing, 65°C)	72°C (extension)
40	0.5–1.5 min	2 min	2–10 min

19. Add 1 μl 10 mM deoxyribonucleotides and 2 U *Taq* polymerase at the end of the final cycle and incubate at 72°C for an additional 30 min.

20. Slow cool the reaction to room temperature.

21. To determine if the desired cDNAs were amplified, separate an aliquot of the PCR reaction by agarose gel electrophoresis, carry out Southern blot analysis with a probe from the *lacZ* portion of the amplified fusion transcript. It is useful to know the size of the fusion transcript and therefore the size of the expected cDNA products.[b]

22. Gel purify the largest PCR products that hybridize with the probe following digestion with the appropriate restriction enzymes, and clone the fragments into suitable DNA plasmid vector.

[a] This step removes the uncreacted oligo number 1, nucleotides, and hydrolysed RNA which are undesirable in the subsequent tailing reaction.

[b] For example, if the fusion transcript contains a large fragment from the endogenous gene, (i.e. greater than 1–2 kb), then a heterogeneous smear of amplified products is likely to be seen on the gel.

Acknowledgements

We thank all colleagues who contributed unpublished data on vectors and methods and helped to improve this manuscript, notably Greg Barsh for the FDG staining procedure, Bernhard Herrmann for his protocols on usage of large genomic probes, William Skarnes for his replica plating and RACE protocols, and Rosa Beddington for *Figure 4*. Last but not least we thank Alexandra Joyner for her advice on the manuscript and many helpful comments and criticism.

References

1. O'Kane, C. J. and Gehring, W. J. (1986). *Proc. Natl Acad. Sci. U.S.A.*, **84,** 9123.
2. Bellen, H. J., O'Kane, C. J., Wilson, C., Grossniklaus, U., Kurth Pearson, R., and Gehring, W. J. (1989). *Genes Dev.*, **3,** 1288.
3. Wilson, C., Kurth Pearson, R., Bellen, H. J., O'Kane, C. J., Grossniklaus, U., and Gehring, W. J. (1989). *Genes Dev.*, **3,** 1301.
4. Kothary, R., Clapoff, S., Brown, A., Campbell, R., Peterson, A., and Rossant, R. (1988). *Nature,* **335,** 435.
5. Allen, N. D., Cran, D. G., Barton, S. C., Hettle, S. Reik, W., and Surani, M. A. (1988). *Nature,* **333,** 852.

6. Gossler, A., Joyner, A. L., Rossant, J., and Skarnes, W. C. (1989). *Science,* **244,** 463.
7. Friedrich, G. and Soriano, P. (1991). *Genes Dev.,* **5,** 1513.
8. Skarnes, W. C., Auerbach, A., and Joyner, A. L. (1992). *Genes Dev.,* **6,** 903.
9. Hope, I. A. (1991). *Development,* **113,** 399.
10. von Melchner, H. and Ruley, H. E. (1989). *J. Virol.,* **63,** 3227.
11. von Melchner, H., Reddy, S., and Ruley, H. E. (1990). *Proc. Natl Acad. Sci. U.S.A.,* **87,** 3733.
12. Reddy, S., DeGregory, J. V., von Melchner, H., and Ruley, H. E. (1991). *J. Virol.,* **65,** 1507.
13. von Melchner, H., DeGregori, J. V., Rayburn, H., Reddy, S., Friedel, C., and Ruley, H. E. (1992). *Genes Dev.,* **6,** 919.
14. Korn, R., Schoor, M., Neuhaus, H., Henseling, U., Soininen, R., Zachgo, J., and Gossler, A. (1992). *Mech. Dev.*, **39**, 95.
15. Soininen, R., Schoor, M., Henseling, U., Tepe, C., Kisters-Woike, B., Rossant, J., and Gossler, A. (1992). *Mech. Dev.*, **39**, 111.
16. Perry, M. D. and Moran, L. A. (1987). *Proc. Natl Acad. Sci. U.S.A.,* **84,** 156.
17. Wagner, E. F., Stewart, T. A., and Mince, B. (1981). *Proc. Natl Acad. Sci. U.S.A.,* **78,** 5016.
18. Bhat, K., McBurney, M. W., and Hamada, H. (1988). *Mol. Cell. Biol.*, **8,** 3251.
19. Hamada, H. (1986). *Mol. Cell. Biol.*, **6,** 4179.
20. Hamada, H. (1986). *Mol. Cell. Biol.*, **6,** 4185.
21. Cepko, C. L., Roberts, B. E., and Mulligan, R. C. (1984). *Cell,* **37,** 1053.
22. Seed, B. (1983). *Nucleic Acids Res.,* **11,** 2427.
23. Sambrook, J., Fritsch, E. F., and Maniatis, T. (ed.) (1989). *Molecular cloning: a laboratory manual* (2nd edition). Cold Spring Harbor Laboratory Press, Cold Spring Harbor, N.Y.
24. Brenner, D. G., Lin-Chao, S., and Cohen, S. N. (1989). *Proc. Natl Acad. Sci. U.S.A.,* **86,** 5517.
25. Rüther, U. and Müller-Hill, B. (1983). *EMBO J.,* **2,** 1791.
26. Reiss, B., Sprengel, R., and Schaller, H. (1984). *EMBO J.,* **3,** 3317.
27. King, W., Patel, M. D., Lobel, L. I., Goff, S. P., and Nguyen Huu, M. C. (1985). *Science,* **288,** 554.
28. Macejak, D. G. and Sarnow, P. (1991). *Nature,* **353,** 90.
29. Raetz, C. R. H., Wermuth, M. M., McIntyre, T. M., Esko, J. D., and Wing, D. C. (1982). *Proc. Natl Acad. Sci. U.S.A.,* **79,** 3223.
30. Gal, S. (1987). In *Methods Enzymol.* (ed. M. M. Gottesman), Vol. 151, pp. 104–112. Academic Press, NY.
31. Udy, G. B., Wilson, V. A., and Evans, M. J. (1990). *Technique,* **2,** 88.
32. Nolan, G. P., Fiering, S., Nicolas, J. F., and Herzenberg, L. A. (1988). *Proc. Natl Acad. Sci. U.S.A.,* **85,** 2603.
33. Ochman, H., Gerber, A. S., and Hartl, D. L. (1988). *Genetics,* **120,** 621.
34. Triglia, T., Peterson, M. G., and Kemp, D. J. (1988). *Nucleic Acids Res.,* **16,** 8186.
35. Kaiser, K. and Murray, N. E. (1988). In *DNA cloning: a practical approach* (ed. D. M. Glover), Vol. 1, pp. 1–48. IRL Press, Oxford.
36. Stoker, A. W. (1990). *Nucleic Acids Res.,* **18,** 4290.
37. Innis, M. A. and Gelfand, D. H. (1990). In *PCR protocols* (ed. M. A. Innis, D. H. Gelfand, J. J. Sninsky, and T. J. White), pp. 3–12. Academic Press.

38. Frohman, M. A., Dush, M. K., and Martin, G. R. (1988). *Proc. Natl Acad. Sci. U.S.A.,* **85,** 8998.
39. Kerr, W. G., Nolan, G. P., Serafini, A. T., and Herzenberg, L. A. (1989). *CSHSQB,* **LIV,** 767.

A1

Suppliers of specialist items

A/S Nunc, Kamstrupvej 90, Kamstrup, DK-4000 Roskilde, Denmark.

Arenberg Sage Inc., 57 Cornwall Street, Jamaica Plain, MA 02130, USA.

Beaudouin, 1 et 3 rue Rataud, Paris 5^{e}, France.

BioRad Laboratories, 3300 Regatta Boulevard, Richmond, California 94804, USA.

Boehringer Mannheim GmbH, Biochemica, Sandhofer Strasse 116, Postfach 31 01 29, Germany.

Carolina Biological, #74-2300, 2700 York Road, Burlington, NC 27215, USA.

Clay Adams, Division of Becton Dickson and Co., Parsippany, NJ 07064, USA.

Collaborative Research Inc., Two Oak Park, Bedford, MA 01730, USA.

Costar/Nuclepore, One Alewife Center, Cambridge, MA 02140, USA.

David Kopf Instruments, 7324 Elmo Street, PO Box 636, Tijuna, CA 91042-0636, USA.

Fine Science Tools Inc., 202–277 Mountain Highway, North Vancouver, BC, V7J 2P2, Canada.

Fluka Chemical Corp., PO Box 14508, St Louis, MO 63178-9916, USA.

Gen-Probe, San Diego, CA 92123, USA.

Gibco BRL, 8400 Helgerman Court, Gaithersburg, MD 20877, USA.

HyClone Laboratories Inc., 1725 South HyClone Road, Logan, UT 84321-6212, USA.

Kulzer GmbH Bereich Technik, Philipp-Reis-Strasse 8, D-6393 Wehrheim, Germany.

Miles Laboratories Inc., 77 Belfield Road, Rexdale, Ontario, M9W 1G6, Canada.

Millipore Canada Inc., 3688 Nashua Drive, Mississauga, Ontario, L4V 1M5, Canada.

Molecular Probes Inc., 4849 Pitchford Avenue, Eugene, Oregon 97402, USA.

Otto Hiller Co., PO Box 1294, Madison, WI 53701, USA.

Parco Scientific Co., Instrument Division, 316 Youngstown-Kingsville Road, PO Box 189, Vienna, OH 44473, USA.

Parke Davis, Warner Lambert, Morris Plains, NJ, USA.

Perkin Elmer (Cetus) Canada Inc., 10 Meter Drive, Rexdale, Ontario, M9W 1A4, Canada.

Pharmacia LKB GmbH, Geschaftsbereich Biotechnologie, Munzinger Strasse 9, Postfach 5480, 7800 Freiburg 1, Germany.

Reichart Jung, Cambridge Instruments GmbH, Postfach 120 Heidelbergstrasse 17-19, D-6907, Nusslock, Germany.

Riedel-de Flaeu, Atetienpesellschaft, Wunstrofer Strasse, D-3016 Seelze 1, Germany.

Roboz Surgical Instrument Co. Inc., 9210 Corporate Boulevard, Suite 220, Rockville, MD 20850, USA.

Sarstedt Inc., PO Box 468, Newton, NC 28656-0468, USA.

Schein JE SPT 313, Dental Supply Co. of New England, 80 Fargo Street, Boston, MA 02210, USA.

Sensortek Inc., 154 Huron Avenue, Clifton, NJ 07013, USA.

Sigma Chemical Company, PO Box 14508, St Louis, MO 63178-9916, USA.

Stoeltin Co., 1350 South Kostner Avenue, Chicago, IL 60623, USA.

Syva Canada Inc., 36 Steacie Drive, Kanata, Ontario, K2K 2A9, Canada.

Index

Index

ORDER OTHER TITLES OF INTEREST TODAY

Price list for: UK, Europe, Rest of World (excluding US and Canada)

Forthcoming Titles

124. Human Genetic Disease Analysis Davies, K.E. (Ed)
.... Spiralbound hardback 0-19-963309-6 **£30.00**
.... Paperback 0-19-963308-8 **£18.50**
123. Protein Phosphorylation Hardie, G. (Ed)
.... Spiralbound hardback 0-19-963306-1 **£32.50**
.... Paperback 0-19-963305-3 **£22.50**
122. Immunocytochemistry Beesley, J. (Ed)
.... Spiralbound hardback 0-19-963270-7 **£32.50**
.... Paperback 0-19-963269-3 **£22.50**
121. Tumour Immunobiology Gallagher, G., Rees, R.C. & others (Eds)
.... Spiralbound hardback 0-19-963370-3 **£35.00**
.... Paperback 0-19-963369-X **£25.00**
120. Transcription Factors Latchman, D.S. (Ed)
.... Spiralbound hardback 0-19-963342-8 **£30.00**
.... Paperback 0-19-963341-X **£19.50**
119. Growth Factors McKay, I.A. & Leigh, I. (Eds)
.... Spiralbound hardback 0-19-963360-6 **£30.00**
.... Paperback 0-19-963359-2 **£19.50**
118. Histocompatibility Testing Dyer, P. & Middleton, D. (Eds)
.... Spiralbound hardback 0-19-963364-9 **£32.50**
.... Paperback 0-19-963363-0 **£22.50**
117. Gene Transcription Hames, D.B. & Higgins, S.J. (Eds)
.... Spiralbound hardback 0-19-963292-8 **£35.00**
.... Paperback 0-19-963291-X **£25.00**
116. Electrophysiology Wallis, D.I. (Ed)
.... Spiralbound hardback 0-19-963348-7 **£32.50**
.... Paperback 0-19-963347-9 **£22.50**
115. Biological Data Analysis Fry, J.C. (Ed)
.... Spiralbound hardback 0-19-963340-1 **£50.00**
.... Paperback 0-19-963339-8 **£27.50**
114. Experimental Neuroanatomy Bolam, J.P. (Ed)
.... Spiralbound hardback 0-19-963326-6 **£32.50**
.... Paperback 0-19-963325-8 **£22.50**
112. Lipid Analysis Hamilton, R.J. & Hamilton, S.J. (Eds)
.... Spiralbound hardback 0-19-963098-4 **£35.00**
.... Paperback 0-19-963099-2 **£25.00**
111. Haemopoiesis Testa, N.G. & Molineux, G. (Eds)
.... Spiralbound hardback 0-19-963366-5 **£32.50**
.... Paperback 0-19-963365-7 **£22.50**

Published Titles

113. Preparative Centrifugation Rickwood, D. (Ed)
.... Spiralbound hardback 0-19-963208-1 **£45.00**
.... Paperback 0-19-963211-1 **£25.00**
110. Pollination Ecology Dafni, A.
.... Spiralbound hardback 0-19-963299-5 **£32.50**
.... Paperback 0-19-963298-7 **£22.50**
109. In Situ Hybridization Wilkinson, D.G. (Ed)
.... Spiralbound hardback 0-19-963328-2 **£30.00**
.... Paperback 0-19-963327-4 **£18.50**
108. Protein Engineering Rees, A.R., Sternberg, M.J.E. & others (Eds)
.... Spiralbound hardback 0-19-963139-5 **£35.00**
.... Paperback 0-19-963138-7 **£25.00**
107. Cell-Cell Interactions Stevenson, B.R., Gallin, W.J. & others (Eds)
.... Spiralbound hardback 0-19-963319-3 **£32.50**
.... Paperback 0-19-963318-5 **£22.50**
106. Diagnostic Molecular Pathology: Volume I Herrington, C.S. & McGee, J. O'D. (Eds)
.... Spiralbound hardback 0-19-963237-5 **£30.00**
.... Paperback 0-19-963236-7 **£19.50**
105. Biomechanics-Materials Vincent, J.F.V. (Ed)
.... Spiralbound hardback 0-19-963223-5 **£35.00**
.... Paperback 0-19-963222-7 **£25.00**
104. Animal Cell Culture (2/e) Freshney, R.I. (Ed)
.... Spiralbound hardback 0-19-963212-X **£30.00**
.... Paperback 0-19-963213-8 **£19.50**
103. Molecular Plant Pathology: Volume II Gurr, S.J., McPherson, M.J. & others (Eds)
.... Spiralbound hardback 0-19-963352-5 **£32.50**
.... Paperback 0-19-963351-7 **£22.50**
101. Protein Targeting Magee, A.I. & Wileman, T. (Eds)
.... Spiralbound hardback 0-19-963206-5 **£32.50**
.... Paperback 0-19-963210-3 **£22.50**
100. Diagnostic Molecular Pathology: Volume II: Cell and Tissue Genotyping Herrington, C.S. & McGee, J.O'D. (Eds)
.... Spiralbound hardback 0-19-963239-1 **£30.00**
.... Paperback 0-19-963238-3 **£19.50**
99. Neuronal Cell Lines Wood, J.N. (Ed)
.... Spiralbound hardback 0-19-963346-0 **£32.50**
.... Paperback 0-19-963345-2 **£22.50**
98. Neural Transplantation Dunnett, S.B. & Björklund, A. (Eds)
.... Spiralbound hardback 0-19-963286-3 **£30.00**
.... Paperback 0-19-963285-5 **£19.50**
97. Human Cytogenetics: Volume II: Malignancy and Acquired Abnormalities (2/e) Rooney, D.E. & Czepulkowski, B.H. (Eds)
.... Spiralbound hardback 0-19-963290-1 **£30.00**
.... Paperback 0-19-963289-8 **£22.50**
96. Human Cytogenetics: Volume I: Constitutional Analysis (2/e) Rooney, D.E. & Czepulkowski, B.H. (Eds)
.... Spiralbound hardback 0-19-963288-X **£30.00**
.... Paperback 0-19-963287-1 **£22.50**
95. Lipid Modification of Proteins Hooper, N.M. & Turner, A.J. (Eds)
.... Spiralbound hardback 0-19-963274-X **£32.50**
.... Paperback 0-19-963273-1 **£22.50**
94. Biomechanics-Structures and Systems Biewener, A.A. (Ed)
.... Spiralbound hardback 0-19-963268-5 **£42.50**
.... Paperback 0-19-963267-7 **£25.00**
93. Lipoprotein Analysis Converse, C.A. & Skinner, E.R. (Eds)
.... Spiralbound hardback 0-19-963192-1 **£30.00**
.... Paperback 0-19-963231-6 **£19.50**
92. Receptor-Ligand Interactions Hulme, E.C. (Ed)
.... Spiralbound hardback 0-19-963090-9 **£35.00**
.... Paperback 0-19-963091-7 **£25.00**
91. Molecular Genetic Analysis of Populations Hoelzel, A.R. (Ed)
.... Spiralbound hardback 0-19-963278-2 **£32.50**
.... Paperback 0-19-963277-4 **£22.50**

90. Enzyme Assays Eisenthal, R. & Danson, M.J. (Eds)
.... Spiralbound hardback 0-19-963142-5 **£35.00**
.... Paperback 0-19-963143-3 **£25.00**

89. Microcomputers in Biochemistry Bryce, C.F.A. (Ed)
.... Spiralbound hardback 0-19-963253-7 **£30.00**
.... Paperback 0-19-963252-9 **£19.50**

88. The Cytoskeleton Carraway, K.L. & Carraway, C.A.C. (Eds)
.... Spiralbound hardback 0-19-963257-X **£30.00**
.... Paperback 0-19-963256-1 **£19.50**

87. Monitoring Neuronal Activity Stamford, J.A. (Ed)
.... Spiralbound hardback 0-19-963244-8 **£30.00**
.... Paperback 0-19-963243-X **£19.50**

86. Crystallization of Nucleic Acids and Proteins Ducruix, A. & Gieg‹130›, R. (Eds)
.... Spiralbound hardback 0-19-963245-6 **£35.00**
.... Paperback 0-19-963246-4 **£25.00**

85. Molecular Plant Pathology: Volume I Gurr, S.J., McPherson, M.J. & others (Eds)
.... Spiralbound hardback 0-19-963103-4 **£30.00**
.... Paperback 0-19-963102-6 **£19.50**

84. Anaerobic Microbiology Levett, P.N. (Ed)
.... Spiralbound hardback 0-19-963204-9 **£32.50**
.... Paperback 0-19-963262-6 **£22.50**

83. Oligonucleotides and Analogues Eckstein, F. (Ed)
.... Spiralbound hardback 0-19-963280-4 **£32.50**
.... Paperback 0-19-963279-0 **£22.50**

82. Electron Microscopy in Biology Harris, R. (Ed)
.... Spiralbound hardback 0-19-963219-7 **£32.50**
.... Paperback 0-19-963215-4 **£22.50**

81. Essential Molecular Biology: Volume II Brown, T.A. (Ed)
.... Spiralbound hardback 0-19-963112-3 **£32.50**
.... Paperback 0-19-963113-1 **£22.50**

80. Cellular Calcium McCormack, J.G. & Cobbold, P.H. (Eds)
.... Spiralbound hardback 0-19-963131-X **£35.00**
.... Paperback 0-19-963130-1 **£25.00**

79. Protein Architecture Lesk, A.M.
.... Spiralbound hardback 0-19-963054-2 **£32.50**
.... Paperback 0-19-963055-0 **£22.50**

78. Cellular Neurobiology Chad, J. & Wheal, H. (Eds)
.... Spiralbound hardback 0-19-963106-9 **£32.50**
.... Paperback 0-19-963107-7 **£22.50**

77. PCR McPherson, M.J., Quirke, P. & others (Eds)
.... Spiralbound hardback 0-19-963226-X **£30.00**
.... Paperback 0-19-963196-4 **£19.50**

76. Mammalian Cell Biotechnology Butler, M. (Ed)
.... Spiralbound hardback 0-19-963207-3 **£30.00**
.... Paperback 0-19-963209-X **£19.50**

75. Cytokines Balkwill, F.R. (Ed)
.... Spiralbound hardback 0-19-963218-9 **£35.00**
.... Paperback 0-19-963214-6 **£25.00**

74. Molecular Neurobiology Chad, J. & Wheal, H. (Eds)
.... Spiralbound hardback 0-19-963108-5 **£30.00**
.... Paperback 0-19-963109-3 **£19.50**

73. Directed Mutagenesis McPherson, M.J. (Ed)
.... Spiralbound hardback 0-19-963141-7 **£30.00**
.... Paperback 0-19-963140-9 **£19.50**

72. Essential Molecular Biology: Volume I Brown, T.A. (Ed)
.... Spiralbound hardback 0-19-963110-7 **£32.50**
.... Paperback 0-19-963111-5 **£22.50**

71. Peptide Hormone Action Siddle, K. & Hutton, J.C.
.... Spiralbound hardback 0-19-963070-4 **£32.50**
.... Paperback 0-19-963071-2 **£22.50**

70. Peptide Hormone Secretion Hutton, J.C. & Siddle, K. (Eds)
.... Spiralbound hardback 0-19-963068-2 **£35.00**
.... Paperback 0-19-963069-0 **£25.00**

69. Postimplantation Mammalian Embryos Copp, A.J. & Cockroft, D.L. (Eds)
.... Spiralbound hardback 0-19-963088-7 **£35.00**
.... Paperback 0-19-963089-5 **£25.00**

68. Receptor-Effector Coupling Hulme, E.C. (Ed)
.... Spiralbound hardback 0-19-963094-1 **£30.00**
.... Paperback 0-19-963095-X **£19.50**

67. Gel Electrophoresis of Proteins (2/e) Hames, B.D. & Rickwood, D. (Eds)
.... Spiralbound hardback 0-19-963074-7 **£35.00**
.... Paperback 0-19-963075-5 **£25.00**

66. Clinical Immunology Gooi, H.C. & Chapel, H. (Eds)
.... Spiralbound hardback 0-19-963086-0 **£32.50**
.... Paperback 0-19-963087-9 **£22.50**

65. Receptor Biochemistry Hulme, E.C. (Ed)
.... Spiralbound hardback 0-19-963092-5 **£35.00**
.... Paperback 0-19-963093-3 **£25.00**

64. Gel Electrophoresis of Nucleic Acids (2/e) Rickwood, D. & Hames, B.D. (Eds)
.... Spiralbound hardback 0-19-963082-8 **£32.50**
.... Paperback 0-19-963083-6 **£22.50**

63. Animal Virus Pathogenesis Oldstone, M.B.A. (Ed)
.... Spiralbound hardback 0-19-963100-X **£30.00**
.... Paperback 0-19-963101-8 **£18.50**

62. Flow Cytometry Ormerod, M.G. (Ed)
.... Paperback 0-19-963053-4 **£22.50**

61. Radioisotopes in Biology Slater, R.J. (Ed)
.... Spiralbound hardback 0-19-963080-1 **£32.50**
.... Paperback 0-19-963081-X **£22.50**

60. Biosensors Cass, A.E.G. (Ed)
.... Spiralbound hardback 0-19-963046-1 **£30.00**
.... Paperback 0-19-963047-X **£19.50**

59. Ribosomes and Protein Synthesis Spedding, G. (Ed)
.... Spiralbound hardback 0-19-963104-2 **£32.50**
.... Paperback 0-19-963105-0 **£22.50**

58. Liposomes New, R.R.C. (Ed)
.... Spiralbound hardback 0-19-963076-3 **£35.00**
.... Paperback 0-19-963077-1 **£22.50**

57. Fermentation McNeil, B. & Harvey, L.M. (Eds)
.... Spiralbound hardback 0-19-963044-5 **£30.00**
.... Paperback 0-19-963045-3 **£19.50**

56. Protein Purification Applications Harris, E.L.V. & Angal, S. (Eds)
.... Spiralbound hardback 0-19-963022-4 **£30.00**
.... Paperback 0-19-963023-2 **£18.50**

55. Nucleic Acids Sequencing Howe, C.J. & Ward, E.S. (Eds)
.... Spiralbound hardback 0-19-963056-9 **£30.00**
.... Paperback 0-19-963057-7 **£19.50**

54. Protein Purification Methods Harris, E.L.V. & Angal, S. (Eds)
.... Spiralbound hardback 0-19-963002-X **£30.00**
.... Paperback 0-19-963003-8 **£20.00**

53. Solid Phase Peptide Synthesis Atherton, E. & Sheppard, R.C.
.... Spiralbound hardback 0-19-963066-6 **£30.00**
.... Paperback 0-19-963067-4 **£18.50**

52. Medical Bacteriology Hawkey, P.M. & Lewis, D.A. (Eds)
.... Spiralbound hardback 0-19-963008-9 **£38.00**
.... Paperback 0-19-963009-7 **£25.00**

51. Proteolytic Enzymes Beynon, R.J. & Bond, J.S. (Eds)
.... Spiralbound hardback 0-19-963058-5 **£30.00**
.... Paperback 0-19-963059-3 **£19.50**

50. Medical Mycology Evans, E.G.V. & Richardson, M.D. (Eds)
.... Spiralbound hardback 0-19-963010-0 **£37.50**
.... Paperback 0-19-963011-9 **£25.00**

49. Computers in Microbiology Bryant, T.N. & Wimpenny, J.W.T. (Eds)
.... Paperback 0-19-963015-1 **£19.50**

48. Protein Sequencing Findlay, J.B.C. & Geisow, M.J. (Eds)
.... Spiralbound hardback 0-19-963012-7 **£30.00**
.... Paperback 0-19-963013-5 **£18.50**

47. Cell Growth and Division Baserga, R. (Ed)
.... Spiralbound hardback 0-19-963026-7 **£30.00**
.... Paperback 0-19-963027-5 **£18.50**

46. Protein Function Creighton, T.E. (Ed)
.... Spiralbound hardback 0-19-963006-2 **£32.50**
.... Paperback 0-19-963007-0 **£22.50**

45. Protein Structure Creighton, T.E. (Ed)
.... Spiralbound hardback 0-19-963000-3 **£32.50**
.... Paperback 0-19-963001-1 **£22.50**

44. Antibodies: Volume II Catty, D. (Ed)
.... Spiralbound hardback 0-19-963018-6 **£30.00**
.... Paperback 0-19-963019-4 **£19.50**

43. HPLC of Macromolecules Oliver, R.W.A. (Ed)
.... Spiralbound hardback 0-19-963020-8 **£30.00**
.... Paperback 0-19-963021-6 **£19.50**
42. Light Microscopy in Biology Lacey, A.J. (Ed)
.... Spiralbound hardback 0-19-963036-4 **£30.00**
.... Paperback 0-19-963037-2 **£19.50**
41. Plant Molecular Biology Shaw, C.H. (Ed)
.... Paperback 1-85221-056-7 **£22.50**
40. Microcomputers in Physiology Fraser, P.J. (Ed)
.... Spiralbound hardback 1-85221-129-6 **£30.00**
.... Paperback 1-85221-130-X **£19.50**
39. Genome Analysis Davies, K.E. (Ed)
.... Spiralbound hardback 1-85221-109-1 **£30.00**
.... Paperback 1-85221-110-5 **£18.50**
38. Antibodies: Volume I Catty, D. (Ed)
.... Paperback 0-947946-85-3 **£19.50**
37. Yeast Campbell, I. & Duffus, J.H. (Eds)
.... Paperback 0-947946-79-9 **£19.50**
36. Mammalian Development Monk, M. (Ed)
.... Hardback 1-85221-030-3 **£30.50**
.... Paperback 1-85221-029-X **£22.50**
35. Lymphocytes Klaus, G.G.B. (Ed)
.... Hardback 1-85221-018-4 **£30.00**
34. Lymphokines and Interferons Clemens, M.J., Morris, A.G. & others (Eds)
.... Paperback 1-85221-035-4 **£22.50**
33. Mitochondria Darley-Usmar, V.M., Rickwood, D. & others (Eds)
.... Hardback 1-85221-034-6 **£32.50**
.... Paperback 1-85221-033-8 **£22.50**
32. Prostaglandins and Related Substances Benedetto, C., McDonald-Gibson, R.G. & others (Eds)
.... Hardback 1-85221-032-X **£32.50**
.... Paperback 1-85221-031-1 **£22.50**
31. DNA Cloning: Volume III Glover, D.M. (Ed)
.... Hardback 1-85221-049-4 **£30.00**
.... Paperback 1-85221-048-6 **£19.50**
30. Steroid Hormones Green, B. & Leake, R.E. (Eds)
.... Paperback 0-947946-53-5 **£19.50**
29. Neurochemistry Turner, A.J. & Bachelard, H.S. (Eds)
.... Hardback 1-85221-028-1 **£30.00**
.... Paperback 1-85221-027-3 **£19.50**
28. Biological Membranes Findlay, J.B.C. & Evans, W.H. (Eds)
.... Hardback 0-947946-84-5 **£32.50**
.... Paperback 0-947946-83-7 **£22.50**
27. Nucleic Acid and Protein Sequence Analysis Bishop, M.J. & Rawlings, C.J. (Eds)
.... Hardback 1-85221-007-9 **£35.00**
.... Paperback 1-85221-006-0 **£25.00**
26. Electron Microscopy in Molecular Biology Sommerville, J. & Scheer, U. (Eds)
.... Hardback 0-947946-64-0 **£30.00**
.... Paperback 0-947946-54-3 **£19.50**
25. Teratocarcinomas and Embryonic Stem Cells Robertson, E.J. (Ed)
.... Hardback 1-85221-005-2 **£19.50**
.... Paperback 1-85221-004-4 **£19.50**
24. Spectrophotometry and Spectrofluorimetry Harris, D.A. & Bashford, C.L. (Eds)
.... Hardback 0-947946-69-1 **£30.00**
.... Paperback 0-947946-46-2 **£18.50**
23. Plasmids Hardy, K.G. (Ed)
.... Paperback 0-947946-81-0 **£18.50**
22. Biochemical Toxicology Snell, K. & Mullock, B. (Eds)
.... Paperback 0-947946-52-7 **£19.50**
19. Drosophila Roberts, D.B. (Ed)
.... Hardback 0-947946-66-7 **£32.50**
.... Paperback 0-947946-45-4 **£22.50**
17. Photosynthesis: Energy Transduction Hipkins, M.F. & Baker, N.R. (Eds)
.... Hardback 0-947946-63-2 **£30.00**
.... Paperback 0-947946-51-9 **£18.50**
16. Human Genetic Diseases Davies, K.E. (Ed)
.... Hardback 0-947946-76-4 **£30.00**
.... Paperback 0-947946-75-6 **£18.50**
14. Nucleic Acid Hybridisation Hames, B.D. & Higgins, S.J. (Eds)
.... Hardback 0-947946-61-6 **£30.00**
.... Paperback 0-947946-23-3 **£19.50**
13. Immobilised Cells and Enzymes Woodward, J. (Ed)
.... Hardback 0-947946-60-8 **£18.50**
12. Plant Cell Culture Dixon, R.A. (Ed)
.... Paperback 0-947946-22-5 **£19.50**
11a. DNA Cloning: Volume I Glover, D.M. (Ed)
.... Paperback 0-947946-18-7 **£18.50**
11b. DNA Cloning: Volume II Glover, D.M. (Ed)
.... Paperback 0-947946-19-5 **£19.50**
10. Virology Mahy, B.W.J. (Ed)
.... Paperback 0-904147-78-9 **£19.50**
9. Affinity Chromatography Dean, P.D.G., Johnson, W.S. & others (Eds)
.... Paperback 0-904147-71-1 **£19.50**
7. Microcomputers in Biology Ireland, C.R. & Long, S.P. (Eds)
.... Paperback 0-904147-57-6 **£18.00**
6. Oligonucleotide Synthesis Gait, M.J. (Ed)
.... Paperback 0-904147-74-6 **£18.50**
5. Transcription and Translation Hames, B.D. & Higgins, S.J. (Eds)
.... Paperback 0-904147-52-5 **£22.50**
3. Iodinated Density Gradient Media Rickwood, D. (Ed)
.... Paperback 0-904147-51-7 **£19.50**

Sets

Essential Molecular Biology: Volumes I and II as a set Brown, T.A. (Ed)
.... Spiralbound hardback 0-19-963114-X **£58.00**
.... Paperback 0-19-963115-8 **£40.00**
Antibodies: Volumes I and II as a set Catty, D. (Ed)
.... Paperback 0-19-963063-1 **£33.00**
Cellular and Molecular Neurobiology Chad, J. & Wheal, H. (Eds)
.... Spiralbound hardback 0-19-963255-3 **£56.00**
.... Paperback 0-19-963254-5 **£38.00**
Protein Structure and Protein Function: Two-volume set Creighton, T.E. (Ed)
.... Spiralbound hardback 0-19-963064-X **£55.00**
.... Paperback 0-19-963065-8 **£38.00**
DNA Cloning: Volumes I, II, III as a set Glover, D.M. (Ed)
.... Paperback 1-85221-069-9 **£46.00**
Molecular Plant Pathology: Volumes I and II as a set Gurr, S.J., McPherson, M.J. & others (Eds)
.... Spiralbound hardback 0-19-963354-1 **£56.00**
.... Paperback 0-19-963353-3 **£37.00**
Protein Purification Methods, and Protein Purification Applications, two-volume set Harris, E.L.V. & Angal, S. (Eds)
.... Spiralbound hardback 0-19-963048-8 **£48.00**
.... Paperback 0-19-963049-6 **£32.00**
Diagnostic Molecular Pathology: Volumes I and II as a set Herrington, C.S. & McGee, J. O'D. (Eds)
.... Spiralbound hardback 0-19-963241-3 **£54.00**
.... Paperback 0-19-963240-5 **£35.00**
Receptor Biochemistry; Receptor-Effector Coupling; Receptor-Ligand Interactions Hulme, E.C. (Ed)
.... Spiralbound hardback 0-19-963096-8 **£90.00**
.... Paperback 0-19-963097-6 **£62.50**
Signal Transduction Milligan, G. (Ed)
.... Spiralbound hardback 0-19-963296-0 **£30.00**
.... Paperback 0-19-963295-2 **£18.50**
Human Cytogenetics: Volumes I and II as a set (2/e) Rooney, D.E. & Czepulkowski, B.H. (Eds)
.... Hardback 0-19-963314-2 **£58.50**
.... Paperback 0-19-963313-4 **£40.50**
Peptide Hormone Secretion/Peptide Hormone Action Siddle, K. & Hutton, J.C. (Eds)
.... Spiralbound hardback 0-19-963072-0 **£55.00**
.... Paperback 0-19-963073-9 **£38.00**

ORDER FORM for UK, Europe and Rest of World

(Excluding USA and Canada)

Qty	ISBN	Author	Title	Amount
			P&P	
			TOTAL	

Please add postage and packing: £1.75 for UK orders under £20; £2.75 for UK orders over £20; overseas orders add 10% of total.

Name ....

Address ....

....

.... Post code ....

[ ] Please charge £ .... to my credit card

Access/VISA/Eurocard/AMEX/Diners Club (circle appropriate card)

Card No .... Expiry date ....

Signature ....

Credit card account address if different from above:

....

.... Postcode ....

[ ] I enclose a cheque for £.....................

Please return this form to: OUP Distribution Services, Saxon Way West, Corby, Northants NN18 9ES

OR ORDER BY CREDIT CARD HOTLINE: Tel +44-(0)536-741519 or Fax +44-(0)536-746337

ORDER OTHER TITLES OF INTEREST TODAY

Price list for: USA and Canada

123. Protein Phosphorylation Hardie, G. (Ed)
.... Spiralbound hardback 0-19-963306-1 **$65.00**
.... Paperback 0-19-963305-3 **$45.00**

121. Tumour Immunobiology Gallagher, G., Rees, R.C. & others (Eds)
.... Spiralbound hardback 0-19-963370-3 **$72.00**
.... Paperback 0-19-963369-X **$50.00**

117. Gene Transcription Hames, D.B. & Higgins, S.J. (Eds)
.... Spiralbound hardback 0-19-963292-8 **$72.00**
.... Paperback 0-19-963291-X **$50.00**

116. Electrophysiology Wallis, D.I. (Ed)
.... Spiralbound hardback 0-19-963348-7 **$66.50**
.... Paperback 0-19-963347-9 **$45.95**

115. Biological Data Analysis Fry, J.C. (Ed)
.... Spiralbound hardback 0-19-963340-1 **$80.00**
.... Paperback 0-19-963339-8 **$60.00**

114. Experimental Neuroanatomy Bolam, J.P. (Ed)
.... Spiralbound hardback 0-19-963326-6 **$65.00**
.... Paperback 0-19-963325-8 **$40.00**

111. Haemopoiesis Testa, N.G. & Molineux, G. (Eds)
.... Spiralbound hardback 0-19-963366-5 **$65.00**
.... Paperback 0-19-963365-7 **$45.00**

113. Preparative Centrifugation Rickwood, D. (Ed)
.... Spiralbound hardback 0-19-963208-1 **$90.00**
.... Paperback 0-19-963211-1 **$50.00**

110. Pollination Ecology Dafni, A.
.... Spiralbound hardback 0-19-963299-5 **$65.00**
.... Paperback 0-19-963298-7 **$45.00**

109. In Situ Hybridization Wilkinson, D.G. (Ed)
.... Spiralbound hardback 0-19-963328-2 **$58.00**
.... Paperback 0-19-963327-4 **$36.00**

108. Protein Engineering Rees, A.R., Sternberg, M.J.E. & others (Eds)
.... Spiralbound hardback 0-19-963139-5 **$75.00**
.... Paperback 0-19-963138-7 **$50.00**

107. Cell-Cell Interactions Stevenson, B.R., Gallin, W.J. & others (Eds)
.... Spiralbound hardback 0-19-963319-3 **$60.00**
.... Paperback 0-19-963318-5 **$40.00**

106. Diagnostic Molecular Pathology: Volume I Herrington, C.S. & McGee, J. O'D. (Eds)
.... Spiralbound hardback 0-19-963237-5 **$58.00**
.... Paperback 0-19-963236-7 **$38.00**

105. Biomechanics-Materials Vincent, J.F.V. (Ed)
.... Spiralbound hardback 0-19-963223-5 **$70.00**
.... Paperback 0-19-963222-7 **$50.00**

104. Animal Cell Culture (2/e) Freshney, R.I. (Ed)
.... Spiralbound hardback 0-19-963212-X **$60.00**
.... Paperback 0-19-963213-8 **$40.00**

103. Molecular Plant Pathology: Volume II Gurr, S.J., McPherson, M.J. & others (Eds)
.... Spiralbound hardback 0-19-963352-5 **$65.00**
.... Paperback 0-19-963351-7 **$45.00**

101. Protein Targeting Magee, A.I. & Wileman, T. (Eds)
.... Spiralbound hardback 0-19-963206-5 **$75.00**
.... Paperback 0-19-963210-3 **$50.00**

100. Diagnostic Molecular Pathology: Volume II: Cell and Tissue Genotyping Herrington, C.S. & McGee, J.O'D. (Eds)
.... Spiralbound hardback 0-19-963239-1 **$60.00**
.... Paperback 0-19-963238-3 **$39.00**

99. Neuronal Cell Lines Wood, J.N. (Ed)
.... Spiralbound hardback 0-19-963346-0 **$68.00**
.... Paperback 0-19-963345-2 **$48.00**

98. Neural Transplantation Dunnett, S.B. & Björklund, A. (Eds)
.... Spiralbound hardback 0-19-963286-3 **$69.00**
.... Paperback 0-19-963285-5 **$42.00**

97. Human Cytogenetics: Volume II: Malignancy and Acquired Abnormalities (2/e) Rooney, D.E. & Czepulkowski, B.H. (Eds)
.... Spiralbound hardback 0-19-963290-1 **$75.00**
.... Paperback 0-19-963289-8 **$50.00**

96. Human Cytogenetics: Volume I: Constitutional Analysis (2/e) Rooney, D.E. & Czepulkowski, B.H. (Eds)
.... Spiralbound hardback 0-19-963288-X **$75.00**
.... Paperback 0-19-963287-1 **$50.00**

95. Lipid Modification of Proteins Hooper, N.M. & Turner, A.J. (Eds)
.... Spiralbound hardback 0-19-963274-X **$75.00**
.... Paperback 0-19-963273-1 **$50.00**

94. Biomechanics-Structures and Systems Biewener, A.A. (Ed)
.... Spiralbound hardback 0-19-963268-5 **$85.00**
.... Paperback 0-19-963267-7 **$50.00**

93. Lipoprotein Analysis Converse, C.A. & Skinner, E.R. (Eds)
.... Spiralbound hardback 0-19-963192-1 **$65.00**
.... Paperback 0-19-963231-6 **$42.00**

92. Receptor-Ligand Interactions Hulme, E.C. (Ed)
.... Spiralbound hardback 0-19-963090-9 **$75.00**
.... Paperback 0-19-963091-7 **$50.00**

91. Molecular Genetic Analysis of Populations Hoelzel, A.R. (Ed)
.... Spiralbound hardback 0-19-963278-2 **$65.00**
.... Paperback 0-19-963277-4 **$45.00**

90. Enzyme Assays Eisenthal, R. & Danson, M.J. (Eds)
.... Spiralbound hardback 0-19-963142-5 **$68.00**
.... Paperback 0-19-963143-3 **$48.00**

89. Microcomputers in Biochemistry Bryce, C.F.A. (Ed)
.... Spiralbound hardback 0-19-963253-7 **$60.00**
.... Paperback 0-19-963252-9 **$40.00**

88. The Cytoskeleton Carraway, K.L. & Carraway, C.A.C. (Eds)
.... Spiralbound hardback 0-19-963257-X **$60.00**
.... Paperback 0-19-963256-1 **$40.00**

87. Monitoring Neuronal Activity Stamford, J.A. (Ed)
.... Spiralbound hardback 0-19-963244-8 **$60.00**
.... Paperback 0-19-963243-X **$40.00**

86. Crystallization of Nucleic Acids and Proteins Ducruix, A. & Gieg‹130›, R. (Eds)
.... Spiralbound hardback 0-19-963245-6 **$60.00**
.... Paperback 0-19-963246-4 **$50.00**

85. Molecular Plant Pathology: Volume I Gurr, S.J., McPherson, M.J. & others (Eds)
.... Spiralbound hardback 0-19-963103-4 **$60.00**
.... Paperback 0-19-963102-6 **$40.00**

84. Anaerobic Microbiology Levett, P.N. (Ed)
.... Spiralbound hardback 0-19-963204-9 **$75.00**
.... Paperback 0-19-963262-6 **$45.00**

83. **Oligonucleotides and Analogues** Eckstein, F. (Ed)
.... Spiralbound hardback 0-19-963280-4 **$65.00**
.... Paperback 0-19-963279-0 **$45.00**

82. **Electron Microscopy in Biology** Harris, R. (Ed)
.... Spiralbound hardback 0-19-963219-7 **$65.00**
.... Paperback 0-19-963215-4 **$45.00**

81. **Essential Molecular Biology: Volume II** Brown, T.A. (Ed)
.... Spiralbound hardback 0-19-963112-3 **$65.00**
.... Paperback 0-19-963113-1 **$45.00**

80. **Cellular Calcium** McCormack, J.G. & Cobbold, P.H. (Eds)
.... Spiralbound hardback 0-19-963131-X **$75.00**
.... Paperback 0-19-963130-1 **$50.00**

79. **Protein Architecture** Lesk, A.M.
.... Spiralbound hardback 0-19-963054-2 **$65.00**
.... Paperback 0-19-963055-0 **$45.00**

78. **Cellular Neurobiology** Chad, J. & Wheal, H. (Eds)
.... Spiralbound hardback 0-19-963106-9 **$73.00**
.... Paperback 0-19-963107-7 **$43.00**

77. **PCR** McPherson, M.J., Quirke, P. & others (Eds)
.... Spiralbound hardback 0-19-963226-X **$55.00**
.... Paperback 0-19-963196-4 **$40.00**

76. **Mammalian Cell Biotechnology** Butler, M. (Ed)
.... Spiralbound hardback 0-19-963207-3 **$60.00**
.... Paperback 0-19-963209-X **$40.00**

75. **Cytokines** Balkwill, F.R. (Ed)
.... Spiralbound hardback 0-19-963218-9 **$64.00**
.... Paperback 0-19-963214-6 **$44.00**

74. **Molecular Neurobiology** Chad, J. & Wheal, H. (Eds)
.... Spiralbound hardback 0-19-963108-5 **$56.00**
.... Paperback 0-19-963109-3 **$36.00**

73. **Directed Mutagenesis** McPherson, M.J. (Ed)
.... Spiralbound hardback 0-19-963141-7 **$55.00**
.... Paperback 0-19-963140-9 **$35.00**

72. **Essential Molecular Biology: Volume I** Brown, T.A. (Ed)
.... Spiralbound hardback 0-19-963110-7 **$65.00**
.... Paperback 0-19-963111-5 **$45.00**

71. **Peptide Hormone Action** Siddle, K. & Hutton, J.C.
.... Spiralbound hardback 0-19-963070-4 **$70.00**
.... Paperback 0-19-963071-2 **$50.00**

70. **Peptide Hormone Secretion** Hutton, J.C. & Siddle, K. (Eds)
.... Spiralbound hardback 0-19-963068-2 **$70.00**
.... Paperback 0-19-963069-0 **$50.00**

69. **Postimplantation Mammalian Embryos** Copp, A.J. & Cockroft, D.L. (Eds)
.... Spiralbound hardback 0-19-963088-7 **$70.00**
.... Paperback 0-19-963089-5 **$50.00**

68. **Receptor-Effector Coupling** Hulme, E.C. (Ed)
.... Spiralbound hardback 0-19-963094-1 **$70.00**
.... Paperback 0-19-963095-X **$45.00**

67. **Gel Electrophoresis of Proteins (2/e)** Hames, B.D. & Rickwood, D. (Eds)
.... Spiralbound hardback 0-19-963074-7 **$75.00**
.... Paperback 0-19-963075-5 **$50.00**

66. **Clinical Immunology** Gooi, H.C. & Chapel, H. (Eds)
.... Spiralbound hardback 0-19-963086-0 **$69.95**
.... Paperback 0-19-963087-9 **$50.00**

65. **Receptor Biochemistry** Hulme, E.C. (Ed)
.... Spiralbound hardback 0-19-963092-5 **$70.00**
.... Paperback 0-19-963093-3 **$50.00**

64. **Gel Electrophoresis of Nucleic Acids (2/e)** Rickwood, D. & Hames, B.D. (Eds)
.... Spiralbound hardback 0-19-963082-8 **$75.00**
.... Paperback 0-19-963083-6 **$50.00**

63. **Animal Virus Pathogenesis** Oldstone, M.B.A. (Ed)
.... Spiralbound hardback 0-19-963100-X **$68.00**
.... Paperback 0-19-963101-8 **$40.00**

62. **Flow Cytometry** Ormerod, M.G. (Ed)
.... Paperback 0-19-963053-4 **$50.00**

61. **Radioisotopes in Biology** Slater, R.J. (Ed)
.... Spiralbound hardback 0-19-963080-1 **$75.00**
.... Paperback 0-19-963081-X **$45.00**

60. **Biosensors** Cass, A.E.G. (Ed)
.... Spiralbound hardback 0-19-963046-1 **$65.00**
.... Paperback 0-19-963047-X **$43.00**

59. **Ribosomes and Protein Synthesis** Spedding, G. (Ed)
.... Spiralbound hardback 0-19-963104-2 **$75.00**
.... Paperback 0-19-963105-0 **$45.00**

58. **Liposomes** New, R.R.C. (Ed)
.... Spiralbound hardback 0-19-963076-3 **$70.00**
.... Paperback 0-19-963077-1 **$45.00**

57. **Fermentation** McNeil, B. & Harvey, L.M. (Eds)
.... Spiralbound hardback 0-19-963044-5 **$65.00**
.... Paperback 0-19-963045-3 **$39.00**

56. **Protein Purification Applications** Harris, E.L.V. & Angal, S. (Eds)
.... Spiralbound hardback 0-19-963022-4 **$54.00**
.... Paperback 0-19-963023-2 **$36.00**

55. **Nucleic Acids Sequencing** Howe, C.J. & Ward, E.S. (Eds)
.... Spiralbound hardback 0-19-963056-9 **$59.00**
.... Paperback 0-19-963057-7 **$38.00**

54. **Protein Purification Methods** Harris, E.L.V. & Angal, S. (Eds)
.... Spiralbound hardback 0-19-963002-X **$60.00**
.... Paperback 0-19-963003-8 **$40.00**

53. **Solid Phase Peptide Synthesis** Atherton, E. & Sheppard, R.C.
.... Spiralbound hardback 0-19-963066-6 **$58.00**
.... Paperback 0-19-963067-4 **$39.95**

52. **Medical Bacteriology** Hawkey, P.M. & Lewis, D.A. (Eds)
.... Spiralbound hardback 0-19-963008-9 **$69.95**
.... Paperback 0-19-963009-7 **$50.00**

51. **Proteolytic Enzymes** Beynon, R.J. & Bond, J.S. (Eds)
.... Spiralbound hardback 0-19-963058-5 **$60.00**
.... Paperback 0-19-963059-3 **$39.00**

50. **Medical Mycology** Evans, E.G.V. & Richardson, M.D. (Eds)
.... Spiralbound hardback 0-19-963010-0 **$69.95**
.... Paperback 0-19-963011-9 **$50.00**

49. **Computers in Microbiology** Bryant, T.N. & Wimpenny, J.W.T. (Eds)
.... Paperback 0-19-963015-1 **$40.00**

48. **Protein Sequencing** Findlay, J.B.C. & Geisow, M.J. (Eds)
.... Spiralbound hardback 0-19-963012-7 **$56.00**
.... Paperback 0-19-963013-5 **$38.00**

47. **Cell Growth and Division** Baserga, R. (Ed)
.... Spiralbound hardback 0-19-963026-7 **$62.00**
.... Paperback 0-19-963027-5 **$38.00**

46. **Protein Function** Creighton, T.E. (Ed)
.... Spiralbound hardback 0-19-963006-2 **$65.00**
.... Paperback 0-19-963007-0 **$45.00**

45. **Protein Structure** Creighton, T.E. (Ed)
.... Spiralbound hardback 0-19-963000-3 **$65.00**
.... Paperback 0-19-963001-1 **$45.00**

44. **Antibodies: Volume II** Catty, D. (Ed)
.... Spiralbound hardback 0-19-963018-6 **$58.00**
.... Paperback 0-19-963019-4 **$39.00**

43. **HPLC of Macromolecules** Oliver, R.W.A. (Ed)
.... Spiralbound hardback 0-19-963020-8 **$54.00**
.... Paperback 0-19-963021-6 **$45.00**

42. **Light Microscopy in Biology** Lacey, A.J. (Ed)
.... Spiralbound hardback 0-19-963036-4 **$62.00**
.... Paperback 0-19-963037-2 **$38.00**

41. **Plant Molecular Biology** Shaw, C.H. (Ed)
.... Paperback 1-85221-056-7 **$38.00**

40. **Microcomputers in Physiology** Fraser, P.J. (Ed)
.... Spiralbound hardback 1-85221-129-6 **$54.00**
.... Paperback 1-85221-130-X **$36.00**

39. **Genome Analysis** Davies, K.E. (Ed)
.... Spiralbound hardback 1-85221-109-1 **$54.00**
.... Paperback 1-85221-110-5 **$36.00**

38. **Antibodies: Volume I** Catty, D. (Ed)
.... Paperback 0-947946-85-3 **$38.00**

37. **Yeast** Campbell, I. & Duffus, J.H. (Eds)
.... Paperback 0-947946-79-9 **$36.00**

36. **Mammalian Development** Monk, M. (Ed)
.... Hardback 1-85221-030-3 **$60.00**
.... Paperback 1-85221-029-X **$45.00**

35. **Lymphocytes** Klaus, G.G.B. (Ed)
.... Hardback 1-85221-018-4 **$54.00**

34. **Lymphokines and Interferons** Clemens, M.J., Morris, A.G. & others (Eds)
.... Paperback 1-85221-035-4 **$44.00**

33. **Mitochondria** Darley-Usmar, V.M., Rickwood, D. & others (Eds)
.... Hardback 1-85221-034-6 **$65.00**
.... Paperback 1-85221-033-8 **$45.00**

32. **Prostaglandins and Related Substances** Benedetto, C., McDonald-Gibson, R.G. & others (Eds)
.... Hardback 1-85221-032-X **$58.00**
.... Paperback 1-85221-031-1 **$38.00**

31. **DNA Cloning: Volume III** Glover, D.M. (Ed)
.... Hardback 1-85221-049-4 **$56.00**
.... Paperback 1-85221-048-6 **$36.00**

30. **Steroid Hormones** Green, B. & Leake, R.E. (Eds)
.... Paperback 0-947946-53-5 **$40.00**

29. **Neurochemistry** Turner, A.J. & Bachelard, H.S. (Eds)
.... Hardback 1-85221-028-1 **$56.00**
.... Paperback 1-85221-027-3 **$36.00**

28. **Biological Membranes** Findlay, J.B.C. & Evans, W.H. (Eds)
.... Hardback 0-947946-84-5 **$54.00**
.... Paperback 0-947946-83-7 **$36.00**

27. **Nucleic Acid and Protein Sequence Analysis** Bishop, M.J. & Rawlings, C.J. (Eds)
.... Hardback 1-85221-007-9 **$66.00**
.... Paperback 1-85221-006-0 **$44.00**

26. **Electron Microscopy in Molecular Biology** Sommerville, J. & Scheer, U. (Eds)
.... Hardback 0-947946-64-0 **$54.00**
.... Paperback 0-947946-54-3 **$40.00**

25. **Teratocarcinomas and Embryonic Stem Cells** Robertson, E.J. (Ed)
.... Hardback 1-85221-005-2 **$62.00**
.... Paperback 1-85221-004-4 **$0.00**

24. **Spectrophotometry and Spectrofluorimetry** Harris, D.A. & Bashford, C.L. (Eds)
.... Hardback 0-947946-69-1 **$56.00**
.... Paperback 0-947946-46-2 **$39.95**

23. **Plasmids** Hardy, K.G. (Ed)
.... Paperback 0-947946-81-0 **$36.00**

22. **Biochemical Toxicology** Snell, K. & Mullock, B. (Eds)
.... Paperback 0-947946-52-7 **$40.00**

19. **Drosophila** Roberts, D.B. (Ed)
.... Hardback 0-947946-66-7 **$67.50**
.... Paperback 0-947946-45-4 **$46.00**

17. **Photosynthesis: Energy Transduction** Hipkins, M.F. & Baker, N.R. (Eds)
.... Hardback 0-947946-63-2 **$54.00**
.... Paperback 0-947946-51-9 **$36.00**

16. **Human Genetic Diseases** Davies, K.E. (Ed)
.... Hardback 0-947946-76-4 **$60.00**
.... Paperback 0-947946-75-6 **$34.00**

14. **Nucleic Acid Hybridisation** Hames, B.D. & Higgins, S.J. (Eds)
.... Hardback 0-947946-61-6 **$60.00**
.... Paperback 0-947946-23-3 **$36.00**

13. **Immobilised Cells and Enzymes** Woodward, J. (Ed)
.... Hardback 0-947946-60-8 **$0.00**

12. **Plant Cell Culture** Dixon, R.A. (Ed)
.... Paperback 0-947946-22-5 **$36.00**

11a. **DNA Cloning: Volume I** Glover, D.M. (Ed)
.... Paperback 0-947946-18-7 **$36.00**

11b. **DNA Cloning: Volume II** Glover, D.M. (Ed)
.... Paperback 0-947946-19-5 **$36.00**

10. **Virology** Mahy, B.W.J. (Ed)
.... Paperback 0-904147-78-9 **$40.00**

9. **Affinity Chromatography** Dean, P.D.G., Johnson, W.S. & others (Eds)
.... Paperback 0-904147-71-1 **$36.00**

7. **Microcomputers in Biology** Ireland, C.R. & Long, S.P. (Eds)
.... Paperback 0-904147-57-6 **$36.00**

6. **Oligonucleotide Synthesis** Gait, M.J. (Ed)
.... Paperback 0-904147-74-6 **$38.00**

5. **Transcription and Translation** Hames, B.D. & Higgins, S.J. (Eds)
.... Paperback 0-904147-52-5 **$38.00**

3. **Iodinated Density Gradient Media** Rickwood, D. (Ed)
.... Paperback 0-904147-51-7 **$36.00**

Sets

Essential Molecular Biology: Volumes I and II as a set Brown, T.A. (Ed)
.... Spiralbound hardback 0-19-963114-X **$118.00**
.... Paperback 0-19-963115-8 **$78.00**

Antibodies: Volumes I and II as a set Catty, D. (Ed)
.... Paperback 0-19-963063-1 **$70.00**

Cellular and Molecular Neurobiology Chad, J. & Wheal, H. (Eds)
.... Spiralbound hardback 0-19-963255-3 **$133.00**
.... Paperback 0-19-963254-5 **$79.00**

Protein Structure and Protein Function: Two-volume set Creighton, T.E. (Ed)
.... Spiralbound hardback 0-19-963064-X **$114.00**
.... Paperback 0-19-963065-8 **$80.00**

DNA Cloning: Volumes I, II, III as a set Glover, D.M. (Ed)
.... Paperback 1-85221-069-9 **$92.00**

Molecular Plant Pathology: Volumes I and II as a set Gurr, S.J., McPherson, M.J. & others (Eds)
.... Spiralbound hardback 0-19-963354-1 **$0.00**
.... Paperback 0-19-963353-3 **$0.00**

Protein Purification Methods, and Protein Purification Applications, two-volume set Harris, E.L.V. & Angal, S. (Eds)
.... Spiralbound hardback 0-19-963048-8 **$98.00**
.... Paperback 0-19-963049-6 **$68.00**

Diagnostic Molecular Pathology: Volumes I and II as a set Herrington, C.S. & McGee, J. O'D. (Eds)
.... Spiralbound hardback 0-19-963241-3 **$0.00**
.... Paperback 0-19-963240-5 **$0.00**

Receptor Biochemistry; Receptor-Effector Coupling; Receptor-Ligand Interactions Hulme, E.C. (Ed)
.... Spiralbound hardback 0-19-963096-8 **$193.00**
.... Paperback 0-19-963097-6 **$125.00**

Signal Transduction Milligan, G. (Ed)
.... Spiralbound hardback 0-19-963296-0 **$60.00**
.... Paperback 0-19-963295-2 **$38.00**

Human Cytogenetics: Volumes I and II as a set (2/e) Rooney, D.E. & Czepulkowski, B.H. (Eds)
.... Hardback 0-19-963314-2 **$130.00**
.... Paperback 0-19-963313-4 **$90.00**

Peptide Hormone Secretion/Peptide Hormone Action Siddle, K. & Hutton, J.C. (Eds)
.... Spiralbound hardback 0-19-963072-0 **$135.00**
.... Paperback 0-19-963073-9 **$90.00**

ORDER FORM for USA and Canada

Qty	ISBN	Author	Title	Amount
			S&H	
			CA and NC residents add appropriate sales tax	
			TOTAL	

Please add shipping and handling: $2.50 for first book, ($1.00 each book thereafter)

Name ....

Address ....

....

.... Zip ....

[ ] Please charge $ .... to my credit card
Mastercard/VISA/American Express (circle appropriate card)

Acct. .... Expiry date ....

Signature ....

Credit card account address if different from above:

....

.... Zip ....

[ ] I enclose a cheque for $...........

Mail orders to: Order Dept. Oxford University Press, 2001 Evans Road, Cary, NC 27513